SpringerBriefs in Materials

The SpringerBriefs Series in Materials presents highly relevant, concise monographs on a wide range of topics covering fundamental advances and new applications in the field. Areas of interest include topical information on innovative, structural and functional materials and composites as well as fundamental principles, physical properties, materials theory and design. SpringerBriefs present succinct summaries of cutting-edge research and practical applications across a wide spectrum of fields. Featuring compact volumes of 50 to 125 pages, the series covers a range of content from professional to academic. Typical topics might include:

- A timely report of state-of-the-art analytical techniques
- A bridge between new research results, as published in journal articles, and a contextual literature review
- A snapshot of a hot or emerging topic
- An in-depth case study or clinical example
- A presentation of core concepts that students must understand in order to make independent contributions

Briefs are characterized by fast, global electronic dissemination, standard publishing contracts, standardized manuscript preparation and formatting guidelines, and expedited production schedules.

More information about this series at http://www.springer.com/series/10111

Aneeya K. Samantara · Satyajit Ratha

Materials Development for Active/Passive Components of a Supercapacitor

Background, Present Status and Future Perspective

 Springer

Aneeya K. Samantara
Colloids and Materials Chemistry
 Department
CSIR-Institute of Minerals and Materials
 Technology
Bhubaneswar, Odisha
India

and

Academy of Scientific and Innovative
 Research
New Delhi
India

Satyajit Ratha
School of Basic Sciences
Indian Institute of Technology Bhubaneswar
Bhubaneswar, Odisha
India

ISSN 2192-1091   ISSN 2192-1105 (electronic)
SpringerBriefs in Materials
ISBN 978-981-10-7262-8   ISBN 978-981-10-7263-5 (eBook)
https://doi.org/10.1007/978-981-10-7263-5

Library of Congress Control Number: 2017958831

© The Author(s) 2018
This work is subject to copyright. All rights are reserved by the Publisher, whether the whole or part
of the material is concerned, specifically the rights of translation, reprinting, reuse of illustrations,
recitation, broadcasting, reproduction on microfilms or in any other physical way, and transmission
or information storage and retrieval, electronic adaptation, computer software, or by similar or dissimilar
methodology now known or hereafter developed.
The use of general descriptive names, registered names, trademarks, service marks, etc. in this
publication does not imply, even in the absence of a specific statement, that such names are exempt from
the relevant protective laws and regulations and therefore free for general use.
The publisher, the authors and the editors are safe to assume that the advice and information in this
book are believed to be true and accurate at the date of publication. Neither the publisher nor the
authors or the editors give a warranty, express or implied, with respect to the material contained herein or
for any errors or omissions that may have been made. The publisher remains neutral with regard to
jurisdictional claims in published maps and institutional affiliations.

Printed on acid-free paper

This Springer imprint is published by Springer Nature
The registered company is Springer Nature Singapore Pte Ltd.
The registered company address is: 152 Beach Road, #21-01/04 Gateway East, Singapore 189721, Singapore

Preface

This book deals with the fabrication and working principle of electrochemical capacitors taking into account the current energy crisis and requirement of highly efficient and long-term energy storage solution. Besides alternatives like rechargeable batteries, electrochemical capacitors could help in addressing the global remedy for cleaner, greener and safer energy storage. This would not only preserve the already depleting non-renewable energy resources (and the environmental hazards resulting out of their consumption) but also provide much required support for renewable energy harvesting resources (i.e. solar, wind, hydro and geothermal).

Though electrochemical capacitors have extraordinary charge accumulation capabilities in comparison to conventional dielectric capacitors, they lag behind rechargeable batteries when it comes to energy per weight ratio (energy density). Therefore, urgent attention is required for further development and optimisation so as to achieve higher energy density in currently available electrochemical capacitors. To accomplish this objective, the book closely examines various important parameters controlling the fabrication and characterisation of electrochemical capacitors of various forms. From these analyses, readers can investigate the role of key components such as electrode/electrolyte materials, separators, current collector and binders on the energy storage performance of electrochemical capacitors.

The first part of this book furnishes detailed emphasis on background, classification, principle of charge storage in electrochemical capacitors. Rest of the discussion is focussed to the current limitations and possible strategies to overcome the same. This book has been written keeping in mind that the broad readership would be graduate students, academic researchers and industries involved in sustainable energy and growth. No other publication has addressed these areas so comprehensively, and therefore, this book can be considered to be highly original in content, with no competing texts.

Bhubaneswar, India/New Delhi, India   Aneeya K. Samantara
Bhubaneswar, India   Satyajit Ratha

About the Book

This brief deals with various forms of supercapacitors starting from traditional carbon based supercapacitors to advanced next generation hybrid supercapacitors. The primary focus is to investigate the successive evolution in the core components of a typical supercapacitor which will bring significant observations regarding their feasibility and overall impact on the charge storage capacity so as to reach at par with the current battery technology. The authors present a critical review of the current collectors, electrode materials and electrolytic components which have distinctive impact on both the power and energy density of a supercapacitor. Emerging trends in the fabrication of hybrid supercapacitor technology bring together the exceptional power density of a double layer capacitor and energy density of a rechargeable battery, which promises a brighter future for the electrical energy storage system.

Contents

1   **Introduction**   1
      References   6

2   **Historical Background and Present Status of the**
      Supercapacitors   9
      References   10

3   **Components of Supercapacitor**   11
      3.1   Electrode Materials   11
           3.1.1   Carbon Based Electrode Materials   13
           3.1.2   Conducting Polymers (CPs)   15
           3.1.3   Metal Oxides   16
           3.1.4   Metal Nitrides   18
           3.1.5   Composite Materials   19
      3.2   Electrolyte Materials   21
           3.2.1   Liquid Electrolytes   22
           3.2.2   Solid and Quasi-Solid (Gel) Type Electrolytes   25
      3.3   Current Collector   28
      3.4   Binders   29
      3.5   Separators   29
      References   30

4   **Asymmetric and Hybrid Supercapacitor**   41
      4.1   Asymmetric Supercapacitor (ASC)   42
      4.2   Hybrid Supercapacitor   42
      References   45

5   **Trend and Scope Beyond Traditional Supercapacitors**   47

About the Authors

Dr. Aneeya K. Samantara has pursued his Ph.D. at CSIR-Institute of Minerals and Materials Technology, Bhubaneswar, Odisha, India. Before joining the Ph.D., he has completed the Master of Philosophy (M.Phil.) in Chemistry from Utkal University and Master of Science in Advanced Organic Chemistry at Ravenshaw University, Cuttack, Odisha. Dr. Aneeya's research interest includes the synthesis of metal chalcogenides and graphene composites for energy storage and conversion application. To his credit, he has authored 16 peer-reviewed articles in international journals and two book chapters.

Mr. Satyajit Ratha is a Project Assistant at the School of Basic Sciences, Indian Institute of Technology Bhubaneswar, India. Prior to joining IIT Bhubaneswar, he received his Bachelor of Science, First-Class Honours from Utkal University, in 2008 and Master of Science from Ravenshaw University in 2010. Satyajit's research interests include two-dimensional semiconductors, nanostructure synthesis, applications, energy storage devices and supercapacitors. He has authored and co-authored about 16 peer-reviewed articles in international journals.

Chapter 1
Introduction

Global population inflation demands for high energy resources to fit the requirements, especially for the emerging economic regions. A sizeable portion of this energy is essential in the field of heavy/local transportation, industrial production, extensive range of electronic gadgets etc. In this context, the future energy requirement is to be shared mostly by the renewable resources such as solar power, wind power, bioenergy, hydro power etc. Harvesting energy from these renewable resources has shown significant improvement considering the fact that hydro power, wind energy and bioenergy (combined) have shown profound growth in their contribution to the world energy production in recent years (RENEWABLES 2016-GLOBAL STATUS REPORT). Though solar power still have not been able to generate enough energy (only 1% share in global energy production), (RENEWABLES 2016-GLOBAL STATUS REPORT) however, rapidly declining costs of solar photovoltaic (PV) technology have created more opportunity for further optimization of the system. As important is the implementation of these renewable energy sources to support the sustainability of our rapidly depleting natural resources, parallel development of high capacity and efficient devices for the storage of the energy harvested from the renewable sources is urgent. Consumption of these non-renewable energy resources, in an unregulated manner would not only destabilize our natural reserves but also affect our environment drastically by releasing particulates and hazardous gaseous pollutants like CO_2, greenhouse gases, sulphur, nitrogen oxide etc. (Fig. 1.1) (Crippa et al. 2016). In order to curb the exhaustion of our priceless natural reserves (for the sake of our future generations) and to protect our surroundings, immediate focus should be put on the generation of clean energy such as electricity, which will ease out the stress on our already degrading environment and though costly (for now), would bring uncompromised and unparalleled balance/benefit to the living world. The compulsion to combat such rapidly changing climatic conditions is the resultant of unregulated consumption of natural resources combined with their disastrous after effects. Realization of renewable energies is a must in order to tackle such catastrophic changes to our living earth. Though these renewable energy sources produce

© The Author(s) 2018
A. K. Samantara and S. Ratha, *Materials Development for Active/Passive Components of a Supercapacitor*, SpringerBriefs in Materials, https://doi.org/10.1007/978-981-10-7263-5_1

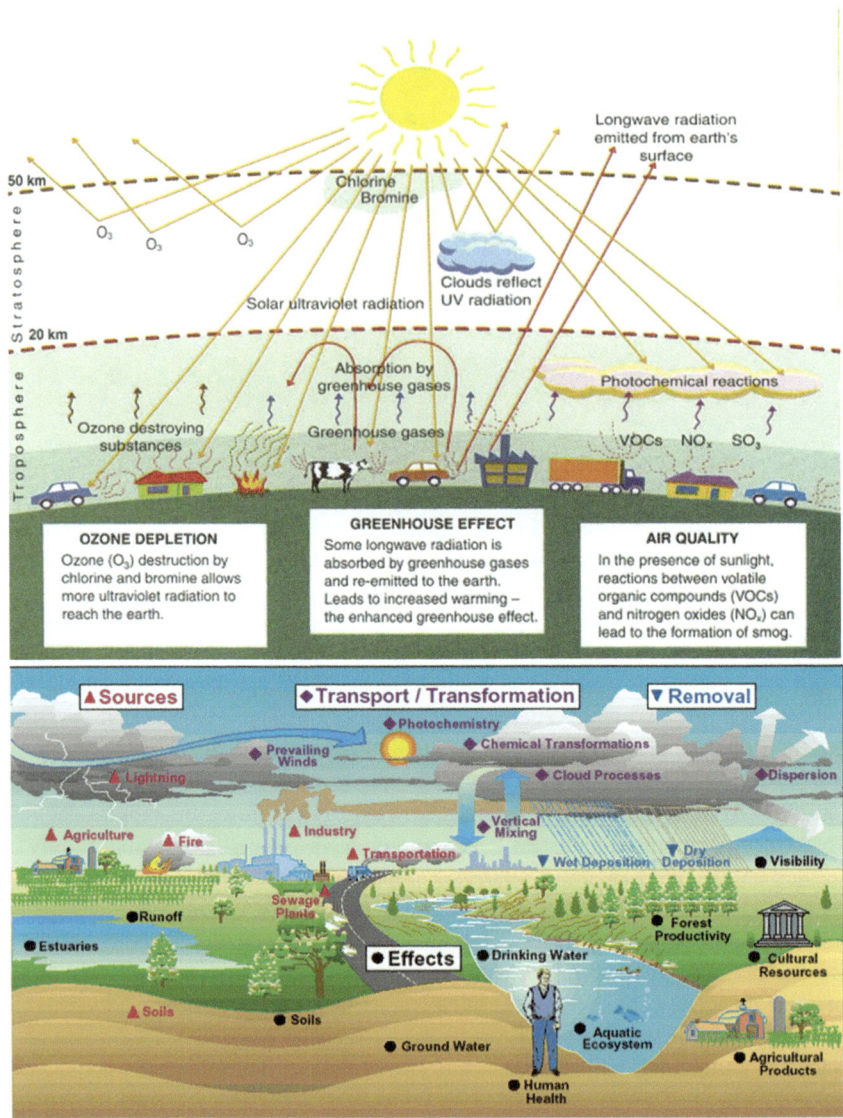

Fig. 1.1 Figure showing the types of air pollutants released due to the consumption of fossil fuels and their direct/indirect consequences on varieties of habitats and their adverse effects (reproduced with permission from Crippa et al. 2016)

enormous amount of energies, they are however, dispersed and thus require localized storage options capable of gathering the enormous bursts of energy generated. Furthermore, the poorly developed storage technologies have long been pushing for the development of efficient, environmentally friendly, cost-effective and safe energy storage options.

Unlike fossil fuels, electrical energy is volatile in nature. Post production, it should immediately be put to carry out effective work and is difficult to store. Therefore, electrical energy storage (EES) requires urgent attention in order to store and disseminate electrical energy properly (Lewis 2007; Yang et al. 2014; Simon et al. 2014). It is to be noted that, electric energy can only be stored via devices like batteries, capacitors and inductors. The battery technology is by far the most evolved and commercially viable option for EES. However, slow charge uptake, bulky design and short cycle life does not allow the battery technology to penetrate critical areas such as high power electric vehicles, lightweight and wearable devices, portable electronics and transparent and flexible displays. Also, hazardous and highly toxic chemical content inside most of these batteries pose great danger to the surrounding environment and require careful management/recycling (Li et al. 2016a). Li-ion technology in recent years has gained prominence due to being highly compact and their low toxic content. Despite the fact that the Li-ion based energy storage shares a sizable portion when it comes to EES, lower values of power density and risk of thermal decomposition at elevated temperature still hinders its implementation in broader context (Jain et al. 2015). Therefore, devices having higher power densities and higher cyclic stability are required to complement the excellent energy density possessed by batteries. The current focus is on the capacitor technology, i.e. to store charge via electrostatic technique. As there are no vigorous chemical reactions taking place, these devices therefore can charge and discharge at lightning speed. Supercapacitors are a new class of devices employing electrostatic mechanism of charge trapping. They possess excellent cyclic stability, ultra-fast charging and discharging and operate through a simple mechanism. Unlike batteries, these devices do not require bulky electrodes and therefore are lightweight, non-toxic, and don't pose complex disposal issues. Supercapacitors possess a capacitance which is $>10^6$ times of the capacitance found in the case of conventional dielectric capacitors. A typical supercapacitor has superior power density than Li-ion battery and higher energy density than a conventional capacitor. Thus, supercapacitor devices can bring a perfect balance between rapid storage of electrical energy and controlled distribution of the same. Therefore, investing a great level of attention into the development and implementation of supercapacitor devices in EES is urgently required.

Besides addressing the issues that were found in the case of Li-ion batteries, supercapacitors can be realized in wider range of applications starting from heavy duty electric vehicles (supporting both the battery stack and effectively replacing the flywheel technology to improve regenerative breaking) to sophisticated low-dimension and portable electronic technologies such as biomechanical energy harvester, flexible and miniature electronic appliances, personal digital assistant (PDA) etc.

The concept of supercapacitors is relatively recent and its practical investigation initiated with the use of carbon based electrodes having exceptionally high surface areas and excellent stability (more than >500,000 operational cycles) (Miller and Burke 2008). Their charge storage is governed by the formation of two distinct layers of charge/ions constituting a double layer. The charge storage is due to the combined effect of both electrochemical adsorptions (electrochemical) and coulombic interaction (electrostatic) at the electrode/electrolyte interface. The charge trapping triggered by adsorption and static attraction effect enables the device to charge and discharge rapidly without losing stability, thus generating more power density than a typical battery.

In the recent past, supercapacitor devices have gone through intense structural and morphological modifications in order to replicate the success story of Li-ion technology. Their morphological appearance and subsequent effective capacitance determines their application in specific field. They can be used in bulk through integration with the battery stacks in heavy duty electric vehicles where they protect the battery units from overcharging and/or deep discharge. They can also be effective in lower-dimensional structures (in-plane supercapacitors) with minimum space requirements (Wu et al. 2013; Li et al. 2016b; Liu et al. 2016, 2017). Charge storage performance is also dependent on the fabrication process. The detailed structural, morphological varieties and effect of both on the supercapacitor performance have been discussed in the following sections.

As mentioned in the previous section, storage of the harvested energy plays a critical role in future high power energy applications, which motivates the development of alternatives for the authentic, cost effective energy storage devices (Largeot et al. 2008; Simon et al. 2014). Although varieties of such alternatives have been considered, the electrochemical energy storage (EES) devices, show greater degree of interest because of their cost effectiveness, scalability and excellent ability to store and deliver energy. Devices like the fuel cells, batteries and supercapacitors etc. are the core components of EES. Although these devices are used for the storage of electrical energy, their fabrication, storage/deliver efficiency and the working principles are different (Gogotsi and Simon 2011; Zhang et al. 2015). Battery and the fuel cell store the electrical energy in terms of chemical reactions, whereas supercapacitors store energy in terms of electric potential developed due to the intercalation/electrosorption of opposite charges under the influence of an external applied field. Conventional supercapacitors store energy via Helmholtz double layer mechanism at the interface of the electrode and electrolyte. While the impressive power density of these devices easily surpasses that of a typical Li-ion battery, their lower value of energy density however, is a primary concern. In order to enhance the energy density without sacrificing the power density, efforts have been made to improve the storage/deliver efficiency of these devices through manipulation of some of the fundamental parameters. Primarily, the specific energy (Wh/kg) and specific power (W/kg) are the two main factors used for the performance evaluation of these storage systems (Fig. 1.2) (Simon and Gogotsi 2008). The capacity of the device has been measured to store the energy and the rate at which the device acquires/deliver energy from/to the supply/load, respectively. Generally, the power and energy of these devices are either normalized with respect to the mass (gravimetric), area (areal) or the volume (volumetric) of the electrode/electrode material.

Capacitors can be classified into three broad categories, (i) Electrostatic, (ii) Electrolytic and (iii) Electrochemical capacitors. Although all three are used to

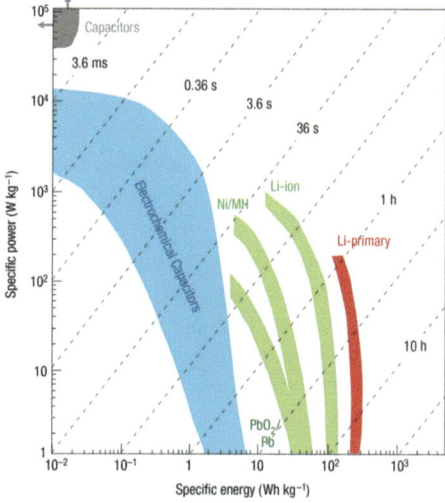

Fig. 1.2 Ragone plots showing representative energy storage devices of capacitors, batteries, and supercapacitors (reproduced with permission from Simon and Gogotsi 2008)

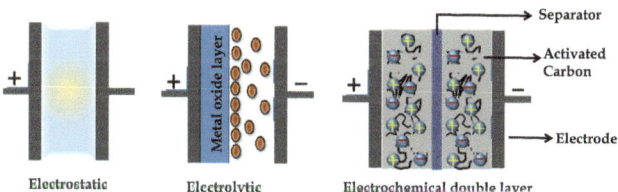

Fig. 1.3 Schematic presentation of the types of capacitors with different configuration

store the stationary electrical energy but they are classified based on their construction and storage mechanism (Fig. 1.3). The electrostatic capacitors are regarded as the first generation of capacitor that contain two metallic plates separated by an insulating (having dielectric property) medium like air, mica sheet, paper, ceramic materials etc. After applying the potential gradient, charges of opposite nature tend to move away from each other and accumulate on the separated electrodes whose polarity is determined by the polarity of the applied field. Capacitance of these capacitors is very low achieving the order of few pico or nano farads. The second generation capacitors are different considering the fact that they have an additional electrolytic content along with the dielectric medium.

It also comprises two electrodes, one of which is a metal foil taken as the positive electrode (anode) and the conducting electrolyte itself acts as the negative electrode (cathode) used to polarize the charges during the charging process. The metal foil (anode) have a thin oxide layer on its surface made through etching process which acts as the dielectric layer and enables to retain the charge for a longer time. The capacitance value of a typical electrolytic capacitor is more than that of an electrostatic capacitor and can reach milli farads. Third generation of capacitors constituted the ceramic type capacitors. These were made from fine granules of ferroelectric materials mixed with ceramic substance to produce desired capacitance. These ceramic type capacitors are the most reliable, robust and have higher capacitance values in comparison to both dielectric and electrolytic capacitors. That is why ceramic capacitors have been implemented in sophisticated electronic circuits as well as military grade applications (Muhammad et al. 2016). It is to be noted that, except electrolytic capacitors, which are polarized, both ceramic type and conventional dielectric capacitors are non-polarized. Also, all of the above mentioned capacitors were not made exclusively for the purpose of energy storage. They were found mainly in the electronic/electrical circuitry for the purpose of coupling, phase shifting, filtering AC (alternative current) signals (ripples) etc.

The next generation capacitors are electrochemical capacitor otherwise known as the electrochemical double layer capacitor (EDLC). In this case, the carbon material is taken as both the cathode and anode separated by an organic or aqueous electrolyte. Here, the charge storage takes place mainly due to the accumulation of the charges/ions on the surface of the electrode material driven by an externally applied potential difference. The capacitance value of these devices are of the order of farads ($>10^{12}$ to 10^9 times the capacitance of a dielectric capacitor and $>10^6$ to 10^3 times that of electrolytic/ceramic capacitor).

References

Crippa M, Janssens-Maenhout G, Dentener F et al (2016) Forty years of improvements in European air quality: regional policy-industry interactions with global impacts. Atmos Chem Phys 16:3825–3841. https://doi.org/10.5194/acp-16-3825-2016

Gogotsi Y, Simon P (2011) True performance metrics in electrochemical energy storage. Science 334(80):917–918

Jain A, Hautier G, Ong SP et al (2015) Relating voltage and thermal safety in Li-ion battery cathodes: a high-throughput computational study. Phys Chem Chem Phys 17:5942–5953. https://doi.org/10.1039/C5CP00250H

Largeot C, Portet C, Chmiola J et al (2008) Relation between the ion size and pore size for an electric double-layer capacitor. J Am Chem Soc 130:2730–2731. https://doi.org/10.1021/ja7106178

Lewis NS (2007) Toward cost-effective solar energy use. Science 315(80):798–801

Li M, Liu J, Han W (2016a) Recycling and management of waste lead-acid batteries: a mini-review. Waste Manag Res 34:298–306. https://doi.org/10.1177/0734242X16633773

Li R-Z, Peng R, Kihm KD et al (2016b) High-rate in-plane micro-supercapacitors scribed onto photo paper using in situ femtolaser-reduced graphene oxide/Au nanoparticle microelectrodes. Energy Environ Sci 9:1458–1467. https://doi.org/10.1039/C5EE03637B

Liu Z, Wu Z-S, Yang S et al (2016) Ultraflexible in-plane micro-supercapacitors by direct printing of solution-processable electrochemically exfoliated graphene. Adv Mater 28:2217–2222. https://doi.org/10.1002/adma.201505304

Liu Z, Liu S, Dong R et al (2017) High power in-plane micro-supercapacitors based on mesoporous polyaniline patterned graphene. Small 13:1603388. https://doi.org/10.1002/smll.201603388

Miller JR, Burke AF (2008) Electrochemical capacitors: challenges and opportunities for real-world applications. Electrochem Soc Interface 17:53–57

Muhammad R, Iqbal Y, Reaney IM (2016) $BaTiO_3$–$Bi(Mg_{2/3}Nb_{1/3})O_3$ ceramics for high-temperature capacitor applications. J Am Ceram Soc 99:2089–2095. https://doi.org/10.1111/jace.14212

Simon P, Gogotsi Y (2008) Materials for electrochemical capacitors. Nat Mater 7:845–854. https://doi.org/10.1038/nmat2297

Simon P, Gogotsi Y, Dunn B (2014) Where do batteries end and supercapacitors begin? Science 343(80):1210–1211. https://doi.org/10.1126/science.1249625

Wu Z, Parvez K, Feng X, Müllen K (2013) Graphene-based in-plane micro-supercapacitors with high power and energy densities. Nat Commun 4:2487

Yang B, Hoober-Burkhardt L, Wang F et al (2014) An inexpensive aqueous flow battery for large-scale electrical energy storage based on water-soluble organic redox couples. J Electrochem Soc 161:A1371–A1380. https://doi.org/10.1149/2.1001409jes

Zhang C, Lv W, Tao Y, Yang Q-H (2015) Towards superior volumetric performance: design and preparation of novel carbon materials for energy storage. Energy Environ Sci 8:1390–1403. https://doi.org/10.1039/C5EE00389J

Chapter 2
Historical Background and Present Status of the Supercapacitors

The capacitive performance of the double layer at solid/electrolyte interface is a known phenomenon since 1879, first predicted by Hermann von Helmholtz. But, the practical use of double layered capacitance was first accomplished after the patent on carbon based electrolytic capacitors booked by Becker at the General Electric Corporation (Becker 1957). After Becker, Sohio Corporation, Cleveland, in the year 1969, manufactured the first electric double layer capacitor for the commercial purpose using porous carbon in a non-aqueous solvent comprising tetra alkyl ammonium salt based electrolyte (Boos 1970). In 1978, NEC (a Japanese multinational Information Technology provider) marketed the double layer capacitor technology as "supercapacitor" for the memory back up in computers. Although the product had been placed in the market, but still it was associated with the lower specific energy values. Thereafter, so many efforts have been made to design supercapacitors with higher efficiency. It has been observed that, up to 1990, no such improvement in the material design and technology development for low cost manufacturing of supercapacitors were carried out. Then Conway's group developed the concept of pseudocapacitance by introducing RuO_2 having higher specific capacitance and low series resistance (Conway 1991). Renewed interest toward the development of supercapacitors occurred after recognizing its importance in the hybrid vehicles in the late 90s. These pseudocapacitive materials (especially the transition metal based oxides, polymers etc.) show ~ 10–100 fold higher energy storage performance in comparison to the EDLCs. This enhanced storage properties are due to the highly reversible redox (faradic) process, which is completely different from that of EDLCs. These types of capacitors involve both the non-faradic (electrical double layer) and faradic (redox) charge storage processes (Zhang et al. 2011). In the present stage, the capacitors available in market constitute both the higher surface area carbon materials (activated carbon) apart from highly redox active RuO_2 (Hu et al. 2006). Because of the high powered supercapacitors, the concept of hybrid energy storage system came into existence that contains supercapacitors integrated either to a fuel cell or a battery.

© The Author(s) 2018
A. K. Samantara and S. Ratha, *Materials Development for Active/Passive Components of a Supercapacitor*, SpringerBriefs in Materials, https://doi.org/10.1007/978-981-10-7263-5_2

References

Becker HI (1957) Low voltage electrolytic capacitor. United States Pat. Off. 2–4

Boos DL (1970) Electrolytic capacitor having carbon paste electrodes. 1–10

Conway BE (1991) Transition from "supercapacitor" to "battery" behavior in electrochemical energy storage. J Electrochem Soc 138:1539–1548. https://doi.org/10.1149/1.2085829

Hu C-C, Chang K-H, Lin M-C, Wu Y-T (2006) Design and tailoring of the nanotubular arrayed architecture of hydrous RuO_2 for next generation supercapacitors. Nano Lett 6:2690–2695. https://doi.org/10.1021/nl061576a

Zhang X, Shi W, Zhu J et al (2011) High-power and high-energy-density flexible pseudocapacitor electrodes made from porous CuO nanobelts and single-walled carbon nanotubes. ACS Nano 5:2013–2019. https://doi.org/10.1021/nn1030719

Chapter 3
Components of Supercapacitor

The components and design of the supercapacitors are similar to the batteries. The components of a supercapacitor device consist of; (i) Electrode material, (ii) Electrolyte material, (iii) Current collector, (iv) Binder and (v) Separators (presented in Fig. 3.1). The electrode and electrolyte materials are regarded as the active component and others are the passive components of the supercapacitor. Normally, a supercapacitor consists of two current collectors/active electrode materials that are separated by a layer of electrolyte or separator.

Although, all the active and passive components have contributions to the storage performance of a supercapacitor, but the electrode material and electrolyte both play a major role.

3.1 Electrode Materials

The electrode materials have a major contribution towards the storage performance of a supercapacitor. Based on the electrode materials, the supercapacitors can be classified into three categories; electrochemical double layered capacitor (EDLC), pseudo capacitors and hybrid capacitors. The classification of the supercapacitors on the basis of electrode materials is illustrated in Fig. 3.2 (Hadjipaschalis et al. 2009). The carbon materials like activated carbon, carbon aerogels, carbon nano tubes (CNTs), graphene etc. show the EDLC behavior. These electrode materials are electrochemically inert and the charge storage takes place only due to the physical accumulation of the charges/ions on the electrode surface. On the other hand, the highly redox active transition metal oxides are under the pseudocapacitive category, where the charge storage takes place because of the physico-chemical adsorption (redox reaction) on the electrode surface (Sarangapani et al. 1996; Simon and Gogotsi 2008; Wang et al. 2009b; Babakhani and Ivey 2010a; Raccichini et al. 2015).

© The Author(s) 2018
A. K. Samantara and S. Ratha, *Materials Development for Active/Passive Components of a Supercapacitor*, SpringerBriefs in Materials,
https://doi.org/10.1007/978-981-10-7263-5_3

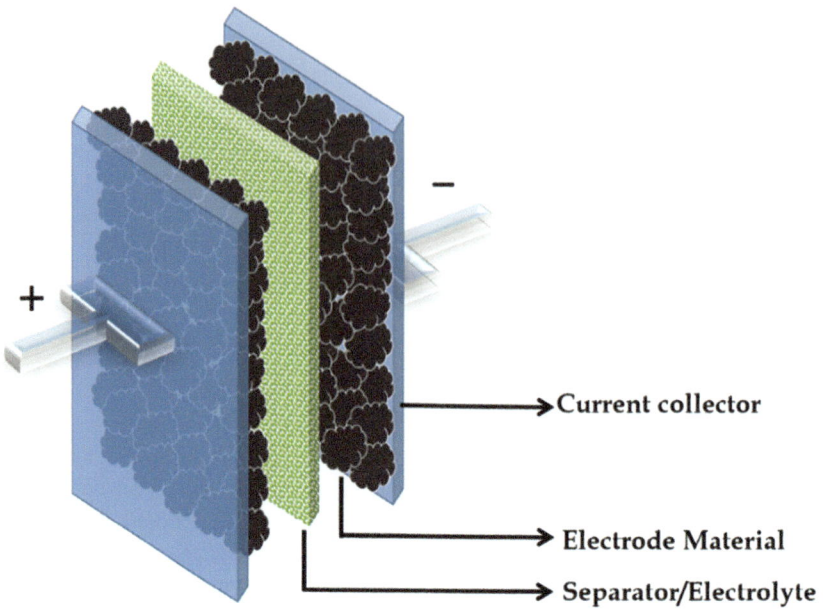

Fig. 3.1 Components of supercapacitor

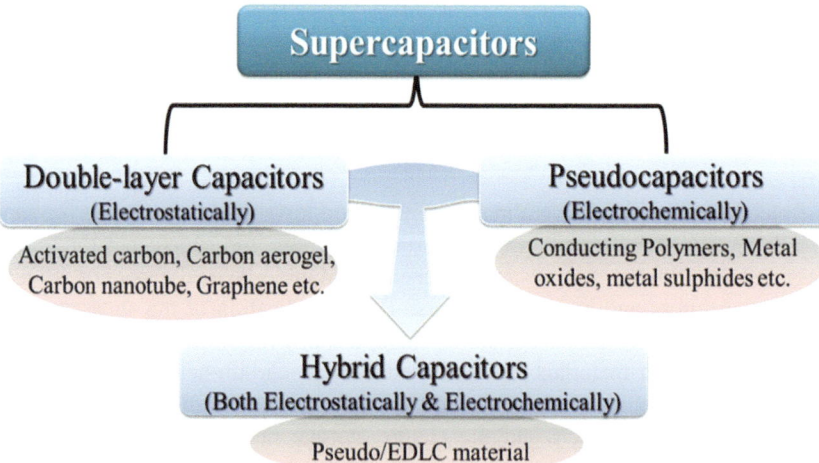

Fig. 3.2 Classification of supercapacitors based on the electrode material and the charge storage mechanism

3.1.1 Carbon Based Electrode Materials

Because of their chemical stability, exceptional electrical conductivity, large theoretical surface area, low manufacture and processing cost, environmental friendliness, high temperature tolerance; carbonaceous materials have gained undivided attention as the electrode material for fabrication of EDLC type supercapacitors (Eikerling et al. 2005; Zhang et al. 2009). The charge storage takes place only by the physical accumulation of charged ions at the electrode/electrolyte interface of the electrode material rather than in the bulk material. Therefore the capacitance performance strongly depends on the specific surface area and surface morphology of the electrode materials. The high surface area carbon materials primarily includes the carbon nanotubes, carbon nano fibers, carbon aerogels, template mediated porous carbon materials etc. (Qu and Shi 1998; Barisci et al. 2000; Endo et al. 2001; Shiraishi et al. 2002; Lee et al. 2006b; Morishita et al. 2006; Fang and Binder 2006; Xu et al. 2008). Many of the researcher have been demonstrated the charge storage performance of these carbon materials and observed the capacitance values of 75–175 F/g and 40–100 F/g in aqueous and organic electrolytes respectively (Burke 2000; Wang et al. 2012a). Generally, the materials having more surface area gives higher capacitance values, thus many methods like heat treatments, steam/alkaline activation, plasma surface treatments etc. have been employed to enhance the specific active surface area of the carbon materials (An et al. 2001; Frackowiak et al. 2002; Raymundo-Piñero et al. 2002, 2006). But in some of the instances, it has been found that the specific capacitance values are not directly proportional to specific surface area of the electrode materials. This is due to the restricted access of the electrolyte ions to all the micro pores in the electrode layer. However, no such concordant information is available on the consequence of optimum pore size of the carbon materials towards the supercapacitor performance. Some of the literatures demonstrated that the optimal pore size may be ~ 0.8–0.4 or ~ 0.7 nm for the organic and aqueous electrolytes respectively (Salitra et al. 2000; Raymundo-Piñero et al. 2006). Meanwhile, along with the surface area and pore size distribution, the surface functional group plays a vital role in the capacitance performance enhancement of carbon materials (Momma et al. 1996; Regisser et al. 1996; Béguin et al. 2005; Hulicova et al. 2005; Frackowiak 2007; Seredych et al. 2008). It has been revealed that the hetero atoms (N, S, O etc.) on the surface of carbon materials increase the adsorption of the electrolyte ions, increasing the hydrophilicity of the carbon materials (Fan et al. 2006; Leitner et al. 2006; Hulicova et al. 2006; Wang et al. 2012a). The wettability of the electrode surface accelerates the transportation of the electrolyte ions throughout the availed micro pores. Therefore, increased charge accumulation on the electrode surface takes place leading to increased specific capacitance values (Wang et al. 2012a). In addition to this, the oxygen containing functional groups on the carbon surface induce redox reactions contributing to 5–10% enhancement in total capacitance value (Salitra et al. 2000; Frackowiak et al. 2002). The higher surface area, porous structure and presence of hetero atoms and functional groups of the carbon material assist a lot to increase the specific capacitance performances. Unfortunately, the high contact resistance between the individual carbon particles increases the series resistance of the electrode, thus the electrochemical supercapacitor performance decreases (Zhang et al. 2009). Therefore the development of more conductive carbon materials for less series resistance is highly indispensable.

Graphene is a well-known 2D allotrope of carbon and is also the building unit of most of the other dimensionalities (Geim and Novoselov 2007). It can be wrapped up into 0-D bucky balls, rolled into 1-D nanotube or stacked into 3-D graphite as can be seen in Fig. 3.3. The high surface area, light weight, layered structure of graphene and presence of different redox sites on the reduced graphene oxide (rGO) surface makes graphene a suitable electrode material for the designing of energy storage devices (Li ion battery, supercapacitor etc.) (Raccichini et al. 2015). Also, because of the theoretically large surface area ($2675 \, m^2/g$) and higher intrinsic specific capacitance value (550 F/g), it has been chosen as the suitable material for electrochemical double layered capacitor (Ivanovskii 2012; Chen et al. 2013; Raccichini et al. 2015). The specific capacitance value of graphene based materials strongly depends upon the synthesis protocol adopted, procedures followed for designing of the electrode, choice of electrolyte etc. (Raccichini et al. 2015).

But the major disadvantage is that during the synthesis of graphene, restacking of the single sheets take place showing decreased value of the theoretical surface area (Raccichini et al. 2015). Moreover the hydrophobic nature restricts free access of the aqueous electrolyte ions all over the surface of graphene leading to lower experimental capacitance values. So, a method to develop hydrophilic graphene with more accessible surface area and less series resistance is highly required.

As discussed in the above sections, the EDLCs (carbon materials) gave limited specific capacitance which is not sufficient for most of the practical applications

Fig. 3.3 2D graphene as the building block of bucky ball, CNTs and graphite. Reproduced with permission from Geim and Novoselov (2007)

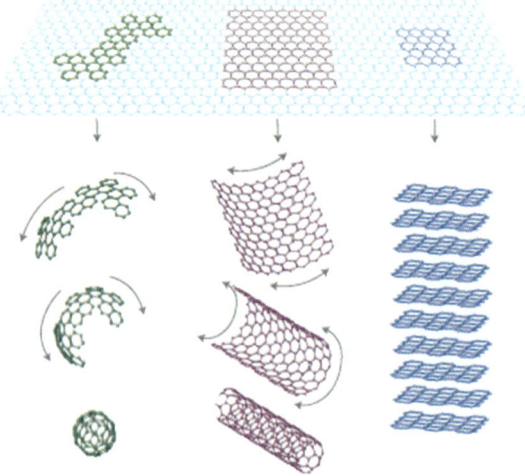

(Zhang et al. 2009). On the other hand, the pseudo capacitive materials (conductive polymers and metal oxides) show 10–100 folds higher specific capacitance compared to the EDLCs because of their fast and reversible faradic reactions (Hu et al. 2007). Thus, substantial research interests have been consecrated to develop the pseudocapacitive electrode materials towards the designing of electrochemical capacitors.

3.1.2 Conducting Polymers (CPs)

The low cost, excellent conductivity (in doped state), negligible environmental impact, wide operational potential window (up to 3.1 V in non-aqueous electrolyte) and tuneable redox activity (by altering the chemical composition) made the conducting polymers (CPs) as the suitable material for supercapacitor development (Kalaji et al. 1999; Prasad et al. 2004; Fan and Maier 2006; Gupta and Miura 2006). The charge storage mechanism in CP based supercapacitors involve the oxidation and reduction of both the surface and bulk of the polymeric backbone of electrode material (Sharma and Bhatti 2010). Depending on the charge developed on CPs during the oxidation or reduction reactions, they are categorised into the following three types; (a) p-doped (PANI; polyaniline, Ppy; polypyrrol etc.), (b) n-doped (polythiophene derivatives) and (c) n-p doped (PTh; polythiophene, poly (3-fluorophenyl) thiophene) CPs (Clemente et al. 1996; Laforgue et al. 1999; Naoi et al. 2000; Mastragostino et al. 2000; Ryu et al. 2002). On the other hand, based on the arrangement of CPs materials, the super capacitors are classified as Type-I (both electrodes are same p-type CP), Type-II (two different p-type CPs) and Type-III (having both n- and p-doped CPs as electrodes) supercapacitors (Rudge et al. 1994; Mastragostino et al. 1996; Conway 1999; Vol'fkovich and Serdyuk 2002; Hashmi and Upadhyaya 2002; Villers et al. 2003). It has been observed that the CPs work well in a restricted potential window. Beyond the confined potential range, either degradation of the CPs takes place (at more positive potential) or the CPs transfers to an insulating state (at more negative potential) (Lota et al. 2004; Snook et al. 2011). Therefore, the selection of a suitable potential window is very much essential for smooth operation of CPs based supercapacitor. Many reports are there on the bare CPs and their composite with carbon materials for supercapacitor electrode development (Lota et al. 2004; Khomenko et al. 2005; Zhang et al. 2008). But unfortunately, they decompose losing the capacitive performance after just about 1000 repeated cycles (Naudin et al. 2001; Sivaraman et al. 2006; Sharma et al. 2008). Also during the intercalation process, swelling and shrinking leads to mechanical degradation of the CPs/CPs composite based supercapacitor electrodes. Therefore, efforts have been devoted to increase the operational stability of these CPs based supercapacitor electrodes. Some research groups have tried to increase the cycle life of the CP based supercapacitors by (i) manipulating the morphology of the CPs, (ii) making hybrid of the polymers, (iii) fabricating the composites with metal oxide/hydroxides/sulphide etc. However, more work still needed to increase the cyclic stability of the CPs based electrode materials towards supercapacitor application (Du Pasquier et al. 2002; Laforgue et al. 2003; Lota et al. 2004; Wang et al. 2006; Sharma et al. 2008).

3.1.3 Metal Oxides

Metal oxides in supercapacitors show better energy storage efficacy over the traditional carbon materials and excellent long cycle performance compared to the CPs based electrode materials. Therefore, efforts have been made to develop higher conductive metal oxides with variable valence states in a cost effective way. Among different transition metal oxides, the likes of RuO_2, MnO_2, cobalt oxides, nickel oxides etc. have been well documented. These oxide based electrode materials stores the electrical charge by both physical accumulation and surface redox reactions in the specified working potential window (Zhao et al. 2007).

Among the transition metal oxides, RuO_2 has been extensively studied for supercapacitor application due to its variable oxidation states, better electrical conductivity, excellent reversibility, robust cyclic stability, wide operational potential window etc. (Sakiyama et al. 1993; Jia et al. 1996; Kim and Kim 2006; Lee et al. 2010a). The charge storage in the RuO_2 based supercapacitors takes place by both the double layer and pseudo capacitive mechanism, demonstrating its candidature for supercapacitor electrode material (Wang et al. 2012a). But the high cost and environmental issues restrict the commercialization of RuO_2 based supercapacitor devices (Fan et al. 2007a). Therefore, other low-cost and environmentally benign metal oxides, e.g. MnO_x, V_2O_5, SnO_2, NiO, CoO, Fe_2O_3 etc. have recently gone under extensive research (Lee and Goodenough 1999; Jayalakshmi et al. 2006; Nam et al. 2008; Wang et al. 2009a; Chen et al. 2009; Wee et al. 2010; Babakhani and Ivey 2010b; Kandalkar et al. 2011). Moreover, the electrochemical performance of the metal oxide based electrode materials towards supercapacitor is highly dependent on the following parameters;

(a) *Specific surface area*: Since the pseudo capacitive performance of the oxide material is a surface phenomenon (i.e. on increasing the surface area, more will be the redox active sites available to interact with the electrolyte), so for higher specific capacitance the specific surface area needs to be increase (Kim et al. 2005; Lee et al. 2006a; Yu et al. 2006). Many effort have been invested to prepare the porous structure materials with a variety of surface morphologies. Among them the one dimensional porous structure are found to be more eminent for the supercapacitor application (Hu et al. 2006; Yuan et al. 2009).

(b) *Surface wettability*: The physically/chemically bound water molecules have a significant role in the enhancement of specific capacitance value in case of oxide based materials. Actually, reversible redox reactions strongly depend on the process involved in the exchange of cations/protons and by electron hopping. It has been observed that, the diffusion of cations taking place by hopping of the electrolyte ions and protons between the OH⁻ and H_2O sites, demonstrating the increased mobility of hydrogen atoms in the hydrated samples compared to the rigid one (Fu et al. 2002). Therefore, the hydrated electrode materials show enhanced cation diffusion inside the electrode layer, leading to more capacitive performance (Liu et al. 1997; Doubova et al. 2004). During the synthesis process, the high temperature annealing results the lower conductive oxide materials. It strongly inhibits the proton intercalation and shows poor ionic conductivity, leading to a decreased pseudo capacitive performance (Rajendra Prasad and Miura 2004; Sugimoto et al. 2006; Wen et al. 2009). So, optimization of the annealing temperature for the material synthesis is highly required.

(c) *Crystallinity*: The energy storage ability of the oxide material also depends on the phase or the crystallinity/crystal structure of that particular material. The increased crystallinity makes the oxide material more rigid and compact surface, restricting the diffusion of electrolyte ions into the bulk (Wang et al. 2012a). So the redox performance of the whole material gets inhibited and the lower capacitance results only from the surface reaction. Whereas, in case of amorphous materials, the capacitance comes from both the surface as well as bulk of the electrode material (Zheng et al. 1995). This can be controlled by manipulating the synthesis procedures followed and starting precursors taken for synthesis (Zheng et al. 1995; Pico et al. 2009). For example, in the case of RuO_2, on increasing the crystallinity, the specific capacitance value decreases (Kim and Popov 2002). This may be due to the restricted diffusion of the protons/electrolyte ions to the bulk of the crystallised electrode material. In the case of MnO_x, it exists in four different crystal structures as α, β, γ and δ-MnO_2 (Zhang and Chen 2008; Donne et al. 2010). Out of different crystal structures, the δ-MnO_2 shows better cation intercalation/deintercalation with very negligible structural deformation leading to higher specific capacitance values (Ma et al. 2004; Athouël et al. 2008).

(d) *Particle size*: In comparison to bulk, nanostructure particles show superior electrochemical performances towards the supercapacitor application (Sugimoto et al. 2004, 2006; Hu et al. 2006; Dubal et al. 2010; Xia et al. 2010). This is due to the availability of high accessible surface area and short path length requirement for the diffusion of the ions/electrons. Thus the smaller sized particle shows high gravimetric capacitance values with more utilization of electrode material surface (Kim and Popov 2002; Pico et al. 2009). The particle size of the oxide materials can be easily manipulated by following the controlled reaction pathways selecting suitable starting precursors (Ramani et al. 2001; Kim and Popov 2002). For example, the specific capacitance of RuO_2 can be brought closer to the theoretical value in its nanoscale dimensions (Sugimoto et al. 2003, 2005).

In view of the above parameters, extensive efforts have been devoted to synthesize a variety of metal oxide nanostructures with a huge number of surface morphologies for the development of supercapacitor electrodes. However, it has been found that the lower electrical conductivity, high consumption of the precious metal (like Ru, Ni etc.), difficulty to control the grain size further proved too much of a challenge for the development of an alternative material. It has been observed that, on introduction of different metals into one of the selected metal oxide can affect the particle size (addition of niobium, vanadium and tin helps to decrease the size of RuO_2 particles), enhance the electrical conductivity (the electrical conductivity of NiO_x increases by inserting the cobalt/molybdenum ions into the NiO_x matrix) and lower the consumption of the precious metal component (introduction of niobium, tin etc. into the RuO_2 matrix decreases the consumption of Ru metal) leading to an advanced electrode material for supercapacitor application (Ramesh et al. 2005; Yokoshima et al. 2006; Fan et al. 2007b; Hu et al. 2007; Brumbach et al. 2010; Wang et al. 2012a; Ratha et al. 2017). Moreover, in some instances it has been found that, during the synthesis of single/binary metal oxides, aggregation of the particles take place and is associated with lower electrical conductivity values. In order to counter these issues, the composite with conductive carbon materials (CNTs, activated carbon, graphene, reduced graphene oxide etc.) have been developed and explored their energy storage performances (Lin et al. 1999; Jeong and Manthiram 2001; Panić et al. 2003; Pico et al. 2008). Fortunately, the highly dispersion of oxide particles was observed with improved access to the whole nanoparticle surface reducing the ionic resistance, thus experienced with enhanced specific capacitance values (Kim et al. 2005; Barranco et al. 2009). Hence, many electrode materials have been developed for designing the supercapacitors. But different materials are associated with their own limitations which are very important to be addressed. The higher specific surface area, porous structure and better conductivity of the carbon materials make them a suitable candidate for the production of commercial supercapacitor devices, but till now they are associated with low specific capacitance values. The as-developed conductive polymers shows higher specific capacitance values, but their shrinking/swelling during the operation leads to poor cycle life. On the other hand, although the amorphous RuO_2 shows better performances towards the supercapacitor application, but the high cost, limited availability and environmental issues restrict its commercial production.

3.1.4 Metal Nitrides

In contrast to metal oxides, enhanced conductivity has been reported in the case of metal nitrides which are emerging as high performance pseudocapacitive materials for asymmetric supercapacitor devices. Due to the high electrical conductivities in comparison to metal oxides, metal nitrides are expected to possess improved power density (Bouhtiyya et al. 2012; Balogun et al. 2015; Das et al. 2015). Lithiated metal nitrides (Li_3N), transition metal nitrides (Mo_2N, VN, TiN) have already shown intriguing electrochemical activities (Bouhtiyya et al. 2012; Balogun et al. 2015). Though application of these metal nitrides is comparatively limited in supercapacitor devices, their robust chemical properties and excellent lithiation/ delithiation property could effectively implemented for both asymmetric and hybrid supercapacitor technologies. In comparison to pseudocapacitive metal oxides, they are capable of generating superior power density. Among the mentioned nitrides, titanium nitride (TiN) has been the most widely studied anode due to its high electrical conductivity and mechanical stability (Grigoras et al. 2016; Kim et al. 2016). However, irreversible oxidation in aqueous media limits its implementation in practical devices. Despite of it, in recent reports, TiN has been combined with more stable materials like graphene, TiO_2, VN, MnO_2, and conducting polymers to form core–shell structures to improve its cycle stability, prevent undesired decomposition to form oxides and rate performances (Balogun et al. 2015).

3.1.5 Composite Materials

As discussed in the previous section, among other carbon based electrode materials, graphene has comparatively higher theoretical specific capacitance and surface area (Raccichini et al. 2015). But it is too difficult to achieve such high values in practice (Sugimoto et al. 2003; Stankovich et al. 2006). Therefore, composite materials of graphene (reduced graphene oxide) with the carbon nanomaterials (carbon nanotubes, activated carbon, carbon spheres, carbon black etc.), metal oxides (MnO_2, RuO_2, Fe_2O_3, Nb_2O_5, NiO etc.) and conducting polymers have been developed (Wu et al. 2010a, b; Yan et al. 2010; Liu et al. 2010; Chen et al. 2011b; Xia et al. 2011; Zhang et al. 2011; Guo and Li 2011; Byon et al. 2011; Yu et al. 2011; Kong et al. 2014, 2015; Grover et al. 2015; Lehtimäki et al. 2015). These additives prevent the restacking process in the reduced graphene oxide generating availability of more accessible surface area to the electrolyte in electrochemical supercapacitors, thus increasing the specific capacitance values. But the complex procedure and requirement of specific instrumental set ups for the synthesis of these composites require intense research and optimization in order to create more opportunities for the development of efficient and high performance electrode materials.

Besides the electrode materials discussed in the previous sections, another class of electrode material consisting of tiny carbon nanoparticles i.e. carbon quantum dots (CQDs) have drawn significant attention in recent past (Xu et al. 2004). These are the quasi spherical carbon particles having very interesting optical performances with tunable electronic properties (Luo et al. 2013; Wang and Hu 2014). These can be synthesized from a variety of precursors by following two main approaches (Fig. 3.4), i.e.

(i) Top-down approach
(ii) Bottom-up approach

In case of the Top-down approach, the quantum dots are derived by breakdown of larger carbonaceous materials. It comprises many of the synthetic strategies like solvothermal/hydrothermal processes, chemical oxidation, arc discharge,

Fig. 3.4 Schematic showing the approaches for synthesis of CQDs

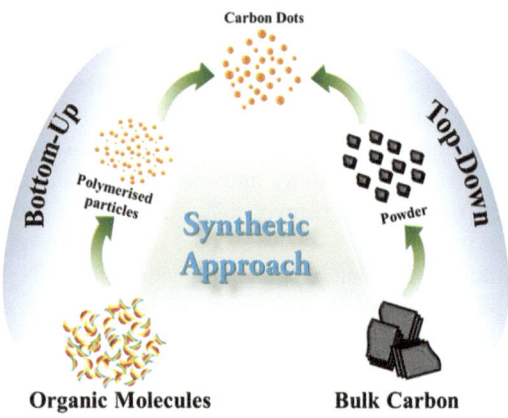

electrochemical cutting, plasma treatment etc. and produces the CQDs from bulk carbon precursors like graphene, carbon fiber and so on (Xu et al. 2004; Pan et al. 2010; Zhu et al. 2011; Shinde and Pillai 2012; Wang et al. 2012b; Li et al. 2012; Peng et al. 2012). Whereas, small carbon containing molecules like carbohydrates, proteins, citric acid etc. produce the CQDs by following the solvothermal/ hydrothermal carbonization, microwave treatment, chemical vapor deposition methods under the bottom-up synthetic approach (Hsu and Chang 2012; Zhai et al. 2012; Tang et al. 2012; Shen et al. 2013; Samantara et al. 2016). The electronic properties of these CQDs can be easily tuned either by inserting the hetero atoms into the carbon skeleton during the synthesis or by the post surface modification of the CQDs (Wang et al. 2010; Yang et al. 2012; Prasad et al. 2013; Sun et al. 2013). Because of the ease of synthesis in bulk quantities without any environmental issues and tunable electronic properties, the CQDs are chosen as the suitable carbon material for the composite preparation with graphene. That will restricts the restacking of individual graphene layers, thus giving enhanced capacitance performance. Thereafter the scalable synthesis of the rGO/CQDs composite in a facile synthetic method and the electrochemical analysis for supercapacitor application is highly desired. So a challenge is there to,

- To enhance the specific capacitance of the carbon materials (by optimising the reaction conditions using suitable starting precursors)
- To choose a low cost starting precursor and a facile synthetic method for the bulk synthesis of the CQDs
- To develop binary metal oxides having better electrical conductivity with porous structure (Augustyn et al. 2014)
- Designing of one dimensional porous nanostructures (due to the reduced diffusion path and more accessible surface area the 1D materials shows better supercapacitor performances)
- To synthesize the composite of metal oxides/metal nitrides with carbon material based electrodes for future supercapacitor devices.

3.2 Electrolyte Materials

The primary challenge in the supercapacitor development is to enhance the energy density value, which can be obtained by either increasing the capacitance of electrode material or by widening the working potential of the device. Ideally, broadening of potential window will be advantageous for increasing the energy density of the device (this is due to the fact that the energy density is a function of square of working potential window) and also in reducing the stacking numbers to reach higher potential differences. Increment in the cell voltage is thus more effective compared to increasing the capacitance of electrode materials for energy density enhancement. As the cell potential strongly depends on the electrochemical stability of electrolyte used in order to avoid exhaustive/parasitic reactions, thus efforts have been made to select a suitable electrolyte material for the supercapacitor development. As a result, a variety of electrolytes have been developed and used in supercapacitors with variable potential windows; aqueous electrolyte based supercapacitors (1.0–1.3 V), organic electrolytes based supercapacitors (2.5–2.7 V), ionic liquid supercapacitors (3.5–4.0 V) etc. The selection of an ideal electrolyte not only enhances the energy density values but also has a significant role in improving cycle life, minimization of internal resistance, increase in the specific power output, control over the operational temperature and to decrease the self-discharge process. A brief classification chart has been illustrated in Fig. 3.5.

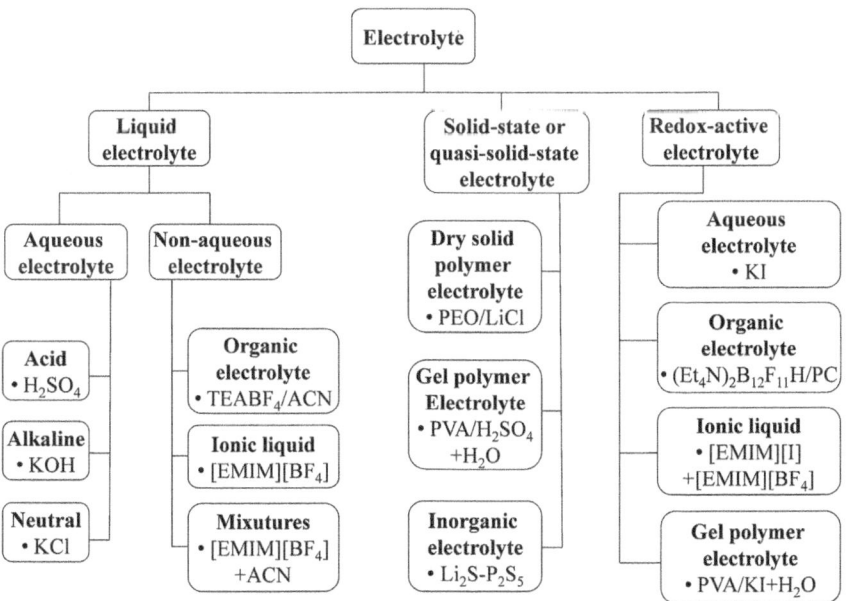

Fig. 3.5 Various categories of electrolytes that have been implemented in the fabrication of supercapacitor devices. Reproduced with permission from Zhong et al. (2015a)

Supercapacitors are developed keeping in mind that they would be able to provide an alternative to Li-ion batteries, thus the quality and quantity of the electrolytic material has a significant role play in tuning a supercapacitor device to obtain desirable output. Preferably, the electrolyte should be lightweight, have good electrochemical activity, high electrical conductivity and large tolerance towards decomposition/degradation. There are varieties of electrolytes that are being employed in supercapacitor devices to validate their overall impact on the capacitive performance. They can be classified into the following broad categories.

3.2.1 Liquid Electrolytes

Choice of electrolytic material for supercapacitor device is the most essential part which is why wide investigation has been done on a large number of liquid electrolytic materials. These liquid electrolytes can further be categorized into aqueous and non-aqueous electrolytes. Further the non-aqueous electrolytes can be categorized into two groups,

1. Organic electrolytes and
2. Ionic liquids.

3.2.1.1 Aqueous Electrolytes

These electrolytes are basically aqueous solution of a group of salts and several inorganic compounds. Aqueous solutions are highly conducting (due to availability of solvated/separated ions) in nature and are beneficial for enhancing the power density (due to low electrolytic resistance resulting in decreased ESR value) in a non-hazardous and cost-effective method. Ease of handling and simplified preparation procedure make the aqueous electrolytes suitable for supercapacitor devices (Zhong et al. 2015a). The aqueous electrolytes are further categorized on the basis of solution pH such as, acidic (H_2SO_4), alkaline (KOH, NaOH, LiOH etc.) and neutral electrolytes (Na_2SO_4, K_2SO_4, Li_2SO_4, $NaNO_3$, KCl etc.) (Long et al. 2011; Fic et al. 2012; Mun et al. 2013; Wang et al. 2014; Dhibar et al. 2014; Abbas et al. 2014; Misnon et al. 2014; Samantara et al. 2015). However, the low potential window, narrow operating temperature range and possible leakage, restricts the use of aqueous electrolyte in commercial supercapacitors (Zhong et al. 2015a). In supercapacitor devices, using aqueous electrolyte, the potential window and operating temperature should be taken care of in order to prevent any kind of decomposition/evaporation which can be catastrophic for the fabricated device. There are instances where it has been found that the neutral electrolyte (aqueous solution of Li_2SO_4) can achieve a wide potential window of 4.0 V and show very low internal resistances in comparison to the organic/ionic liquid based electrolytes (Shimizu et al. 2013). Therefore, the use of aqueous electrolytes in the supercapacitor application might provide a cost effective and environmentally benign route to enhanced energy storage.

3.2.1.2 Non-aqueous Electrolytes

These types of electrolytes are basically organic salt solutions or fused ions (ionic liquids) which are able to sustain higher working potential windows without showing any sign of decomposition/degradation. They are robust in comparison to aqueous electrolytes and are favorable for commercialization purpose. However, the high cost, high toxicity, low conductivity hinders their large scale production. There are three basic categories of non-aqueous electrolytes as discussed in the following Sections.

1. Organic Electrolytes

Organic electrolytes are prepared by dissolving a conducting salt in an organic solvent. Because of the wide potential window (2.5–2.8 V), most of the commercial supercapacitors have been fabricated by using organic electrolytes. This wide cell potential not only provides more energy density but also helps to increase the overall thermal and cyclic stability of as-fabricated supercapacitor devices (Brandt et al. 2012; Jung et al. 2013; Perricone et al. 2013; Zheng et al. 2014). Many of the works on organic electrolyte based supercapacitors make use of tetraethyl ammonium tetrafluoroborate (TEABF$_4$), spiro-(1,1′)-bipyrrolidinium tetrafluoroborate (SBPBF$_4$), triethylmethylammonium tetrafluoroborate (TEMABF$_4$) etc. in acetonitrile/propylene carbonate solvents (Perricone et al. 2013; Sevilla and Fuertes 2014; Zheng et al. 2014). However, supercapacitor devices fabricated using organic electrolytes are of high cost, having lower specific capacitance, higher internal resistance, low ionic conductivity, highly flammable, are toxic and require sophisticated instrumentations for their fabrication. Also, in some instances the degradation in capacitance value and self-discharge issues have been observed (Conway 1999). Besides that, the EDLC performance in these types of electrolytes is considerably poor due to large solvated ions and low dielectric constants. Thus, improvement of the electrolyte material is an essential task for high power/energy electrochemical supercapacitor development.

2. Ionic Liquids (ILs)

Ionic liquids are made of salts that have room temperature melting point (<100 °C) and contain an asymmetric organic cation coordinated with an inorganic/organic anion. These ionic liquids are often dissolved in organic solvents or taken in their pure form (solvent free electrolyte) to investigate the overall impact on the performance of a supercapacitor device. These ILs provide ample choice to be designed by selecting different cation-anion pair in order to optimize the working potential window. In many of the literatures, the ionic liquid employed for the supercapacitor fabrication comprises of cations such as phosphonium, sulfonium, pyrrolidinium, ammonium etc. and anions such as tetrafluoroborate (BF$_4^-$), dicyanamide (DCA⁻), bis(trifluoromethanesulfonyl)imide (TFSI⁻) etc. (Zhong et al. 2015a). They have higher ionic conductivities and wider cell voltages (3.5–4.0 V) in comparison to the organic electrolytes (Shi et al. 2014a). Despite having wider potential windows in comparison to aqueous and organic electrolytes, few factors such as low capacitance values, high cost and higher viscosity of the ionic electrolytes restrict their implementation in commercial supercapacitor devices (Xu et al. 2008; Chen et al. 2011a).

3.2.1.3 Redox-Active Aqueous Electrolytes

With conventional supercapacitors relying solely on the redox activity of the electrode materials, it will be interesting if the electrolyte itself can contribute toward the overall capacitive performance of supercapacitor device (Akinwolemiwa et al. 2015). In general, the electrolyte acts as a media for the movement of charge and/or ions between the two electrodes and remains passive without any contribution towards the net capacitance of the device. This is undesired as most of the supercapacitors contain significant amount of liquid electrolyte in order to induce higher pseudocapacitance. Thus, it sounds promising to extract similar redox based capacitance from the electrolyte itself. Hydroquinones, m-phenylenediamine, KI, lignosulfonates, etc. are some of the commonly used compounds which have shown excellent redox behavior as electrolytes (Yan et al. 2014).

Though large pseudocapacitance is obtained via redox activities from the active components, these redox-active electrolytes have the disadvantage of compromising with the rate performance and cycling stability of the supercapacitor devices owing primarily to their deficient electrochemical stability. Besides, a number of electrolytes tend to generate disproportionately high capacitance at the positive electrode with low capacitance at the other electrode, making it difficult to exploit the exceptionally high specific capacitance. Furthermore, some electrolyte systems seem to contribute high electrochemical activity only in acidic media, thus requiring specialized corrosion resistant current collectors, which is undesired considering the high cost associated with the fabrication of such specific current collectors. This significantly hinders the practical implementation of these electrolytes. Therefore, numerous research efforts still should be focused on scalable synthesis of fairly stable and highly effective redox mediators.

3.2.1.4 Redox-Active Non-aqueous Electrolytes

In a quest to achieve elevated cell voltage which would push for a higher energy density value, several non-aqueous electrolytes including organic and IL-based electrolytes have been investigated and reported in the literature. Inclusion of redox-active polyfluorododecaborate cluster ions (i.e., $[B_{12}F_xH_{12-x}]^{2-}$) into an organic mixture solvent containing propylene carbonate (PC) and dimethyl carbonate (DMC) has shown effective pseudocapacitive contribution to the total capacitance of the carbon based supercapacitors (Béguin et al. 2014; Zhong et al. 2015a). Furthermore, tetraethylammonium undecafluorododecaborate (($Et_4N)_2B_{12}F_{11}H$) when combined with a mixture of PC–DMC, was able to provide overcharge protection for supercapacitor devices due to the presence of redox-active anions (Béguin et al. 2014; Zhong et al. 2015a). Similarly for ionic liquid based electrolytes, it has been demonstrated that an addition of 5 wt% 1-ethyl-3-methylimidazolium iodide ([EMIM][I]) into [EMIM][BF_4] ionic liquid could boost the specific capacitance of a supercapacitor device by nearly ~50% when compared to bare [EMIM][BF_4] ionic liquid electrolyte (Béguin et al. 2014; Yan et al. 2014; Zhong et al. 2015a). Also, another ionic liquid electrolyte, N-ethyl-N-methylpyrrolidinium fluorohydrogenate (EMPyr(FH)$_{2.3}$F) could significantly contribute extra specific capacitance to a supercapacitor through the redox reaction (Béguin et al. 2014; Zhong et al. 2015a).

3.2.2 Solid and Quasi-Solid (Gel) Type Electrolytes

Though they tend to promote the redox reaction due to better ionic conductivity, leakage and solution resistance plays a limiting role for liquid type electrolytes to be used in large scale. Also, in the case of wearable electronic devices and other portable electronic devices, implementation of liquid based supercapacitors is highly discouraged. Widespread investigation is thus being carried out for more robust and reliable electrolytes to minimize spill, leakage which will make supercapacitor devices more convenient to be implemented for next generation practical applications. In this context, significant research has been carried out for both solid and quasi-solid (gel) type electrolytes (Liao et al. 2015). These non-liquid based electrolytes are generating greater interest due to rapid miniaturization in the electronic industries. A typical gel type electrolyte is composed of a polymer matrix embedded with a liquid electrolyte, specifically designed to retain its flexibility. The liquid electrolyte may be an aqueous electrolyte, an organic solvent, an ionic liquid or a redox active electrolytic mixture.

A number of polymers such as poly(vinyl alcohol) (PVA), poly(methyl-methacrylate) (PMMA), potassium/sodium salt of polyacrylic acid (potassium/sodium polyacrylate), poly(ether ether ketone) (PEEK), poly(acrylonitrile)-block-poly(ethylene glycol)-block-poly(acrylonitrile) (PAN-b-PEG-b-PAN), and poly(vinylidene fluoride-co-hexafluoropropylene) (PVDF-HFP) etc. have been investigated for their possible application as a suitable polymer matrix so as to form a desired GPE assembly (Zhong et al. 2015a). A class of GPE known as hydrogels is formed when water is used as the plasticizer. In these hydrogel type GPEs, the water molecules are readily trapped by the complex polymer matrices through surface tension and result in interesting three dimensional network structures (Zhong et al. 2015a). Besides water, few organic solvents such as PC,330 EC and DMF331 or their mixtures (e.g., PC–EC,332 PC–EC–DMC,333 and PC–EC331) have also been commonly used as the plasticizers in GPEs. The level of plasticization in a typical GPE is controlled by the percentage mixture of the polymer matrix and the liquid electrolyte content. The advantage of implementing these GPEs in supercapacitor devices is that they allow for the development of flexible and tunable shapes and structures which are significant in the bendable electronic devices. Various flexible, stretchable, and micro supercapacitor devices have been reported using hydrogels extracted from PVA. Hydrogels based on PVA are synthesized by mixing it with aqueous solutions like H_2SO_4 (acidic), KOH (alkaline) and LiCl (neutral) etc. (Ma et al. 2014; Zhong et al. 2015b). The preference towards the use of PVA as the most suitable polymer matrix comes from its excellent hydrophilicity, simple synthesis technique, non-toxic effect and cost-effectiveness. Hydrogels based on several alternative conducting polymers such as polythiophene, polyaniline and polypyrrole are showing interesting electrolytic properties too (Shi et al. 2014b; Huang et al. 2016; Batisse and Raymundo-Piñero 2017).

Similar to GPEs, solid state electrolytes have recently attracted significant attention with the rapidly increasing demand for reliable and stable storage devices in the areas like portable/wearable electronics, micro-electronic devices (where compactness is highly desired), printable and flexible electronic devices etc. Ease of fabrication and packaging of solid electrolyte based supercapacitors gives a clear advantage over traditional supercapacitor devices those running on liquid based electrolytes. This will also minimize the undesired leakage erstwhile common in the case of liquid electrolyte based supercapacitors.

However, major focus has been put on polymer based elctrolytes for the fabrication of symmetric/asymmetric/hybrid supercapacitors. Only a few works has been reported on the possible implementation of inorganic solid materials such as ceramic electrolytes till date. Polymer based electrolytes can further be grouped into three distinct categories such as solid polymer electrolytes (SPEs), gel polymer electrolytes (GPEs) and polyelectrolytes. The gel polymer type electrolytes are often termed as quasi-solid electrolytes due to the presence of liquid content. In contrast, a solid polymer electrolyte consists of a mixture of polymer and inorganic salt (e.g. LiCl) having zero liquid content. The polymer content of a solid polymer electrolyte provides robust framework which supports both liquid (GPEs) and solid (SPEs) contents, however, the ionic conductivity of GPEs are better than those of SPEs. For this reason, GPE-based supercapacitors currently dominate the solid electrolyte-based supercapacitor devices, and studies on dry SPEs are rather limited. The schematic presentation of both SPE and GPE has been provided in Fig. 3.6.

However, depending on their composition and at high operating temperatures, the reliability and stability of GPEs may degrade significantly due to poor mechanical strength and a narrow operative temperature range particularly when water is used as the solvent. Furthermore, the weak mechanical strength of some GPEs is the main concern, as it may lead to internal short circuits, raising safety concerns. As reported in the literature, solid electrolytes have been used in several types of supercapacitors such as EDLCs, pseudocapacitors and hybrid supercapacitors taking different kinds of electrode materials (Liao et al. 2015). Following

a

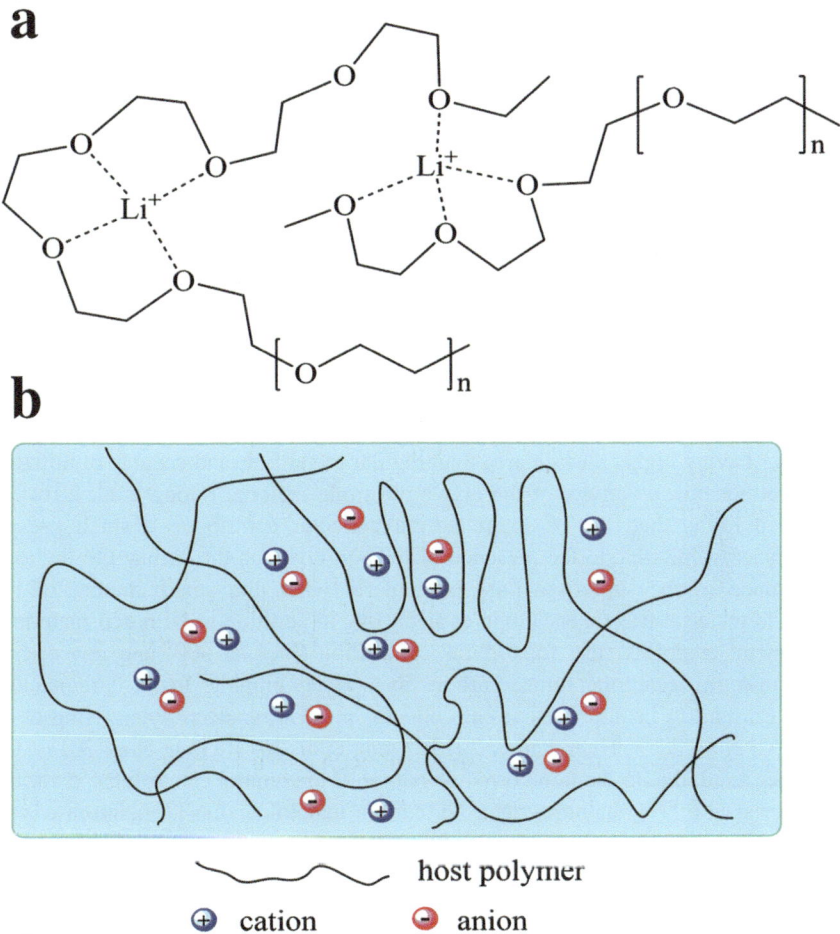

b

Fig. 3.6 Schematic showing **a** a solid polymer electrolyte and **b** gel polymer electrolyte. Reproduced with permission from Zhong et al. (2015a)

criteria should be considered while developing solid electrolytes for supercapacitor devices: (a) high ionic conductivity, (b) high chemical, electrochemical and thermal stability, and (c) sufficient mechanical strength and dimensional stability. In practice, it is difficult for a solid-sate electrolyte to meet all of these requirements. There are often some trade-off between ionic conductivity and mechanical strength. In this aspect, several reviews have been published recently.

3.3   Current Collector

The electrode and electrolyte materials are the active components of supercapacitors, whereas the current collector is a passive component. But, like the electrode/electrolyte materials, it plays a crucial role towards the durability and cell potential enhancement of a supercapacitor device (Zhong et al. 2015a). The selection of a current collector solely depends on the type of electrolyte taken and electrode material with which the supercapacitor device has been fabricated. In case of strong acid based electrolytes, corrosion resistive metal foil current collectors (like Au) have been used. Furthermore, to minimize the cost of supercapacitor devices, materials like ITO plates (indium tin oxide) and carbon based materials have been developed (Cho et al. 2012; Ryu et al. 2014). In the case of alkaline electrolyte based supercapacitors, the low cost nickel (Ni) based materials are chosen as suitable current collectors. Ni foam has gained popularity as excellent current collector having higher surface area than regular metal foils and capable of utilizing the electrochemical activity of the active electrode material (Gong et al. 2014). It has been found that, the Ni foam current collector contributes a small pseudo capacitance value (due to the presence of $NiO/Ni(OH)_x$ on the surface) to the total capacitance of the supercapacitors particularly when the small amount of the electrode material was taken (Gong et al. 2014). In addition to Ni based materials, other metal (stainless steel foils/foams etc.), alloy (Inconel 600) and few carbon based materials (carbon fabrics, carbon fiber paper, graphite foam, carbon cloth etc.) are found to be suitable current collector in alkaline electrolytes (Xing et al. 2011; Xu et al. 2013; Huang et al. 2013; Xiang et al. 2014; Gong et al. 2014). On the other hand, due to the non-corrosive nature of the neutral electrolytes, a variety of material like ITO, stainless steel, Ni, carbon nanotubes (CNT), titanium oxynitride etc. have been used as the current collector for supercapacitors (Xiao et al. 2014; Ratajczak et al. 2014). In the case of non-aqueous electrolyte (organic electrolytes and ionic liquids) based supercapacitors, the aluminum current collectors are broadly used (Bittner et al. 2012; Brandt et al. 2013). Although a variety of current collectors were developed, but each has some advantages and drawbacks and require further investigation.

Selection of both an electrode material and a suitable current collector is not only essential but also critical for the performance of a supercapacitor device. Incompatibility between current collector (indirect fabrication methods such as drop casting, spin coating, vacuum drying, mechanical pressing etc.) significantly affects the capacitive performance due to the equivalent series resistance (ESR). Therefore, in many recent reports, researchers have followed different synthetic techniques so as to grow the desired electrode materials directly on various high conducting, robust and electrochemically inactive current collectors such as Ni-foam, Ni-foil, Cu-foil, graphene, carbon cloth, stainless steel foil etc. Furthermore, both the current collector and electrode material have different properties toward different types of electrolytes. Interaction between the electrode material and electrolyte plays determining factor toward the stability and reliability of a supercapacitor device. Care should also be taken to consider the interface between electrode material and current collector in the presence of electrolytes which will be helpful to optimize/stabilize the whole system.

3.4 Binders

Binders have long been used by mixing with the powdered active electrode material that not only helps to maintain the structural integrity of the electrode material film but also help in achieving better adhesion between the active material and current collector. Generally, fluorinated polymeric materials like, PVDF [poly(vinylidene fluoride)], PTFE (Polytetrafluoroethylene), Nafion, natural cellulose, PVP (polyvinyl pyrrolidone), PAA (polyacrylic acid) and conductive polymers (polypyrrol, polyaniline etc.) have been employed as the binder for the fabrication of supercapacitors (Lee et al. 2010b, 2011; Kang et al. 2014; Varzi et al. 2014; Aslan et al. 2014). It has been observed that the performance of the supercapacitors strongly affected by the content and type of binder used (Tsay et al. 2012; Abbas et al. 2014). However, the presence of binders such as PTFE in excess concentration could have inhibiting effect on the supercapacitor performance due to their hydrophobic property and hinders the electrolyte penetration leading to lower capacitance performance (Tsay et al. 2012). In some of the cases, mixtures of binders in an optimized proportion have been used to get more wettability and electrolyte access. Due to the adverse effects like lower conductivity, reduced active surface and wettability of the electrode material, most of the current supercapacitor technologies are aiming to develop binder free electrodes which would lead to better utilization of the electrochemical activity of the electrode material. The desired adhesion between the electrode material and the current collector has been achieved via direct growth method or by mechanical crimping with the help of hydraulic technique.

3.5 Separators

Similar to current collectors and binders, separator is a passive component of electrochemical capacitors. Although it has no contribution to the capacitive performance, but plays an important role by preventing any physical contact between the electrodes and facilitating electron transfer between them. The separator's properties like (i) electrical insulator, (ii) ion transfer capability, (iii) chemical/ electrochemical inertness, (iv) high mechanical strength (v) optimal thickness, (vi) porosity and (vii) surface morphology etc. influences the performance and durability of a supercapacitor (Tõnurist et al. 2012; Zhong et al. 2015a). There are wide range of materials from which various separators can be fabricated such as, polypropylene, PVDF, PTFE, cellulose polymer membranes (cellulose nitrate, cellulose acetate membranes etc.), glass fiber, Celgard, Nafion 115, graphene oxide films, eggshell membrane etc. (Liu and Pickup 2008; Bittner et al. 2012; Yu et al. 2012; Tõnurist et al. 2012; Shulga et al. 2014).

References

Abbas Q, Pajak D, Frąckowiak E, Béguin F (2014) Effect of binder on the performance of carbon/carbon symmetric capacitors in salt aqueous electrolyte. Electrochim Acta 140:132–138. doi:10.1016/j.electacta.2014.04.096

Akinwolemiwa B, Peng C, Chen GZ (2015) Redox electrolytes in supercapacitors. J Electrochem Soc 162:A5054–A5059. doi:10.1149/2.0111505jes

An KH, Kim WS, Park YS et al (2001) Supercapacitors using single-walled carbon nanotube electrodes. Adv Mater 13:497–500. doi:10.1002/1521-4095(200104)13:7<497::AID-ADMA497>3.0.CO;2-H

Aslan M, Weingarth D, Jäckel N et al (2014) Polyvinylpyrrolidone as binder for castable supercapacitor electrodes with high electrochemical performance in organic electrolytes. J Power Sour 266:374–383. doi:10.1016/j.jpowsour.2014.05.031

Athouël L, Moser F, Dugas R et al (2008) Variation of the MnO$_2$ birnessite structure upon charge/discharge in an electrochemical supercapacitor electrode in aqueous Na$_2$SO$_4$ electrolyte. J Phys Chem C 112:7270–7277. doi:10.1021/jp0773029

Augustyn V, Simon P, Dunn B (2014) Pseudocapacitive oxide materials for high-rate electrochemical energy storage. Energy Environ Sci 7:1597–1614. doi:10.1039/C3EE44164D

Babakhani B, Ivey DG (2010a) Improved capacitive behavior of electrochemically synthesized Mn oxide/PEDOT electrodes utilized as electrochemical capacitors. Electrochim Acta 55:4014–4024. doi:10.1016/j.electacta.2010.02.030

Babakhani B, Ivey DG (2010b) Anodic deposition of manganese oxide electrodes with rod-like structures for application as electrochemical capacitors. J Power Sources 195:2110–2117. doi:10.1016/j.jpowsour.2009.10.045

Balogun M-S, Qiu W, Wang W et al (2015) Recent advances in metal nitrides as high-performance electrode materials for energy storage devices. J Mater Chem A 3:1364–1387. doi:10.1039/C4TA05565A

Barisci JN, Wallace GG, Baughman RH (2000) Electrochemical characterization of single-walled carbon nanotube electrodes. J Electrochem Soc 147:4580–4583. doi:10.1149/1.1394104

Barranco V, Pico F, Ibañez J et al (2009) Amorphous carbon nanofibres inducing high specific capacitance of deposited hydrous ruthenium oxide. Electrochim Acta 54:7452–7457. doi:10.1016/j.electacta.2009.07.080

Batisse N, Raymundo-Piñero E (2017) A self-standing hydrogel neutral electrolyte for high voltage and safe flexible supercapacitors. J Power Sources 348:168–174. doi:10.1016/j.jpowsour.2017.03.005

Béguin F, Szostak K, Lota G, Frackowiak E (2005) A self-supporting electrode for supercapacitors prepared by one-step pyrolysis of carbon nanotube/polyacrylonitrile blends. Adv Mater 17:2380–2384. doi:10.1002/adma.200402103

Béguin F, Presser V, Balducci A, Frackowiak E (2014) Carbons and electrolytes for advanced supercapacitors. Adv Mater 26:2219–2251. doi:10.1002/adma.201304137

Bittner AM, Zhu M, Yang Y et al (2012) Ageing of electrochemical double layer capacitors. J Power Sour 203:262–273. doi:10.1016/j.jpowsour.2011.10.083

Bouhtiyya S, Lucio-Porto R, Ducros J-B et al (2012) Transition metal nitrides thin films for supercapacitor applications. Meet Abstr MA2012-02:494

Brandt A, Isken P, Lex-Balducci A, Balducci A (2012) Adiponitrile-based electrochemical double layer capacitor. J Power Sour 204:213–219. doi:10.1016/j.jpowsour.2011.12.025

Brandt A, Ramirez-Castro C, Anouti M, Balducci A (2013) An investigation about the use of mixtures of sulfonium-based ionic liquids and propylene carbonate as electrolytes for supercapacitors. J Mater Chem A 1:12669–12678. doi:10.1039/C3TA12737K

Brumbach MT, Alam TM, Nilson RH et al (2010) Ruthenium oxide–niobium hydroxide composites for pseudocapacitor electrodes. Mater Chem Phys 124:359–370. doi:10.1016/j.matchemphys.2010.06.047

Burke A (2000) Ultracapacitors: why, how, and where is the technology. J Power Sour 91:37–50. doi:10.1016/S0378-7753(00)00485-7

Byon HR, Lee SW, Chen S et al (2011) Thin films of carbon nanotubes and chemically reduced graphenes for electrochemical micro-capacitors. Carbon N Y 49:457–467. doi:10.1016/j.carbon.2010.09.042

Chen J, Huang K, Liu S (2009) Hydrothermal preparation of octadecahedron Fe_3O_4 thin film for use in an electrochemical supercapacitor. Electrochim Acta 55:1–5. doi:10.1016/j.electacta.2009.04.017

Chen J, Li C, Shi G (2013) Graphene materials for electrochemical capacitors. J Phys Chem Lett 4:1244–1253. doi:10.1021/jz400160k

Chen Y, Zhang X, Zhang D et al (2011a) High performance supercapacitors based on reduced graphene oxide in aqueous and ionic liquid electrolytes. Carbon N Y 49:573–580. doi:10.1016/j.carbon.2010.09.060

Chen Z, Augustyn V, Wen J et al (2011b) High-performance supercapacitors based on intertwined CNT/V_2O_5 nanowire nanocomposites. Adv Mater 23:791–795. doi:10.1002/adma.201003658

Cho HW, Hepowit LR, Nam H-S et al (2012) Synthesis and supercapacitive properties of electrodeposited polyaniline composite electrode on acrylonitrile-butadiene rubber as a flexible current collector. Synth Met 162:410–413. doi:10.1016/j.synthmet.2011.12.025

Clemente A, Panero S, Spila E, Scrosati B (1996) Solid-state, polymer-based, redox capacitors. Solid State Ionics 85:273–277. doi:10.1016/0167-2738(96)00070-7

Conway BE (1999) Electrochemical supercapacitors scientific fundamentals and technological applications. Kluwer Academic/Plenum, New York

Das B, Behm M, Lindbergh G et al (2015) High performance metal nitrides, MN (M = Cr, Co) nanoparticles for non-aqueous hybrid supercapacitors. Adv Powder Technol 26:783–788. doi:10.1016/j.apt.2015.02.001

Dhibar S, Bhattacharya P, Hatui G et al (2014) Transition metal-doped polyaniline/single-walled carbon nanotubes nanocomposites: efficient electrode material for high performance supercapacitors. ACS Sustain Chem Eng 2:1114–1127. doi:10.1021/sc5000072

Donne SW, Hollenkamp AF, Jones BC (2010) Structure, morphology and electrochemical behaviour of manganese oxides prepared by controlled decomposition of permanganate. J Power Sour 195:367–373. doi:10.1016/j.jpowsour.2009.06.103

Doubova LM, De Battisti A, Daolio S et al (2004) Effect of surface structure on behavior of RuO_2 electrodes in sulfuric acid aqueous solution. Russ J Electrochem 40:1115–1122. doi:10.1023/B:RUEL.0000048642.73284.4f

Du Pasquier A, Laforgue A, Simon P et al (2002) A nonaqueous asymmetric hybrid $Li_4Ti_5O_{12}$/poly(fluorophenylthiophene) energy storage device. J Electrochem Soc 149:A302–A306. doi:10.1149/1.1446081

Dubal DP, Dhawale DS, Salunkhe RR, Lokhande CD (2010) Conversion of interlocked cube-like Mn_3O_4 into nanoflakes of layered birnessite MnO_2 during supercapacitive studies. J Alloys Compd 496:370–375. doi:10.1016/j.jallcom.2010.02.014

Eikerling M, Kornyshev AA, Lust E (2005) Optimized structure of nanoporous carbon-based double-layer capacitors. J Electrochem Soc 152:E24–E33. doi:10.1149/1.1825379

Endo M, Maeda T, Takeda T et al (2001) Capacitance and pore-size distribution in aqueous and nonaqueous electrolytes using various activated carbon electrodes. J Electrochem Soc 148:A910–A914. doi:10.1149/1.1382589

Fan L-Z, Maier J (2006) High-performance polypyrrole electrode materials for redox supercapacitors. Electrochem Commun 8:937–940. doi:10.1016/j.elecom.2006.03.035

Fan JY, Wu XL, Li HX et al (2006) Si-based solid blue emitters from 3C-SiC nanocrystals. Appl
Phys A 82:485–487. doi:10.1007/s00339-005-3445-4

Fan L-Z, Hu Y-S, Maier J et al (2007a) High electroactivity of polyaniline in supercapacitors by
using a hierarchically porous carbon monolith as a support. Adv Funct Mater 17:3083–3087.
doi:10.1002/adfm.200700518

Fan Z, Chen J, Cui K et al (2007b) Preparation and capacitive properties of cobalt–nickel oxides/
carbon nanotube composites. Electrochim Acta 52:2959–2965. doi:10.1016/j.electacta.2006.
09.029

Fang B, Binder L (2006) A modified activated carbon aerogel for high-energy storage in electric
double layer capacitors. J Power Sour 163:616–622. doi:10.1016/j.jpowsour.2006.09.014

Fic K, Lota G, Meller M, Frackowiak E (2012) Novel insight into neutral medium as electrolyte
for high-voltage supercapacitors. Energy Environ Sci 5:5842–5850. doi:10.1039/C1EE02262H

Frackowiak E (2007) Carbon materials for supercapacitor application. Phys Chem Chem Phys
9:1774–1785. doi:10.1039/B618139M

Frackowiak E, Delpeux S, Jurewicz K et al (2002) Enhanced capacitance of carbon nanotubes
through chemical activation. Chem Phys Lett 361:35–41. doi:10.1016/S0009-2614(02)00684-X

Fu R, Ma Z, Zheng JP (2002) Proton NMR and dynamic studies of hydrous ruthenium oxide.
J Phys Chem B 106:3592–3596. doi:10.1021/jp013860q

Geim AK, Novoselov KS (2007) The rise of graphene. Nat Mater 6:183–191

Gong X, Cheng JP, Liu F et al (2014) Nickel-Cobalt hydroxide microspheres electrodepositioned
on nickel cobaltite nanowires grown on Ni foam for high-performance pseudocapacitors.
J Power Sour 267:610–616. doi:10.1016/j.jpowsour.2014.05.120

Grigoras K, Keskinen J, Grönberg L et al (2016) Conformal titanium nitride in a porous silicon
matrix: a nanomaterial for in-chip supercapacitors. Nano Energy 26:340–345. doi:10.1016/j.
nanoen.2016.04.029

Grover S, Goel S, Sahu V et al (2015) Asymmetric supercapacitive characteristics of PANI
embedded holey graphene nanoribbons. ACS Sustain Chem Eng 3:1460–1469. doi:10.1021/
acssuschemeng.5b00184

Guo CX, Li CM (2011) A self-assembled hierarchical nanostructure comprising carbon spheres
and graphene nanosheets for enhanced supercapacitor performance. Energy Environ Sci
4:4504–4507. doi:10.1039/C1EE01676H

Gupta V, Miura N (2006) High performance electrochemical supercapacitor from electrochemically synthesized nanostructured polyaniline. Mater Lett 60:1466–1469. doi:10.1016/j.matlet.
2005.11.047

Hadjipaschalis I, Poullikkas A, Efthimiou V (2009) Overview of current and future energy storage
technologies for electric power applications. Renew Sustain Energy Rev 13:1513–1522.
doi:10.1016/j.rser.2008.09.028

Hashmi SA, Upadhyaya HM (2002) Polypyrrole and poly(3-methyl thiophene)-based solid state
redox supercapacitors using ion conducting polymer electrolyte. Solid State Ionics 153:883–
889. doi:10.1016/S0167-2738(02)00390-9

Hsu P-C, Chang H-T (2012) Synthesis of high-quality carbon nanodots from hydrophilic
compounds: role of functional groups. Chem Commun 48:3984–3986. doi:10.1039/
C2CC30188A

Hu C-C, Chang K-H, Lin M-C, Wu Y-T (2006) Design and tailoring of the nanotubular arrayed
architecture of hydrous RuO_2 for next generation supercapacitors. Nano Lett 6:2690–2695.
doi:10.1021/nl061576a

Hu C-C, Wang C-C, Chang K-H (2007) A comparison study of the capacitive behavior for sol–
gel-derived and co-annealed ruthenium–tin oxide composites. Electrochim Acta 52:2691–
2700. doi:10.1016/j.electacta.2006.09.026

Huang L, Chen D, Ding Y et al (2013) Hybrid composite $Ni(OH)_2$@$NiCO_2O_4$ grown on carbon
fiber paper for high-performance supercapacitors. ACS Appl Mater Interfaces 5:11159–11162.
doi:10.1021/am403367u

Huang H, Yao J, Li L et al (2016) Reinforced polyaniline/polyvinyl alcohol conducting hydrogel from a freezing–thawing method as self-supported electrode for supercapacitors. J Mater Sci 51:8728–8736. doi:10.1007/s10853-016-0137-8

Hulicova D, Yamashita J, Soneda Y et al (2005) Supercapacitors prepared from melamine-based carbon. Chem Mater 17:1241–1247. doi:10.1021/cm049337g

Hulicova D, Kodama M, Hatori H (2006) Electrochemical performance of nitrogen-enriched carbons in aqueous and non-aqueous supercapacitors. Chem Mater 18:2318–2326. doi:10.1021/cm060146i

Ivanovskii AL (2012) Graphene-based and graphene-like materials. Russ Chem Rev 81:571–605. doi:10.1070/RC2012v081n07ABEH004302

Jayalakshmi M, Venugopal N, Raja KP, Rao MM (2006) Nano SnO_2–Al_2O_3 mixed oxide and SnO_2–Al_2O_3–carbon composite oxides as new and novel electrodes for supercapacitor applications. J Power Sour 158:1538–1543. doi:10.1016/j.jpowsour.2005.10.091

Jeong YU, Manthiram A (2001) Amorphous tungsten oxide/ruthenium oxide composites for electrochemical capacitors. J Electrochem Soc 148:A189–A193. doi:10.1149/1.1345869

Jia QX, Song SG, Wu XD et al (1996) Epitaxial growth of highly conductive RuO_2 thin films on (100) Si. Appl Phys Lett 68:1069–1071. doi:10.1063/1.115715

Jung N, Kwon S, Lee D et al (2013) Synthesis of chemically bonded graphene/carbon nanotube composites and their application in large volumetric capacitance supercapacitors. Adv Mater 25:6854–6858. doi:10.1002/adma.201302788

Kalaji M, Murphy PJ, Williams GO (1999) The study of conducting polymers for use as redox supercapacitors. Synth Met 102:1360–1361. doi:10.1016/S0379-6779(98)01334-4

Kandalkar SG, Lee H-M, Chae H, Kim C-K (2011) Structural, morphological, and electrical characteristics of the electrodeposited cobalt oxide electrode for supercapacitor applications. Mater Res Bull 46:48–51. doi:10.1016/j.materresbull.2010.09.041

Kang M, Lee JE, Shim HW et al (2014) Intrinsically conductive polymer binders for electrochemical capacitor application. RSC Adv 4:27939–27945. doi:10.1039/C4RA03261F

Khomenko V, Frackowiak E, Béguin F (2005) Determination of the specific capacitance of conducting polymer/nanotubes composite electrodes using different cell configurations. Electrochim Acta 50:2499–2506. doi:10.1016/j.electacta.2004.10.078

Kim H, Popov BN (2002) Characterization of hydrous ruthenium oxide/carbon nanocomposite supercapacitors prepared by a colloidal method. J Power Sour 104:52–61. doi:10.1016/S0378-7753(01)00903-X

Kim I-H, Kim K-B (2006) Electrochemical characterization of hydrous ruthenium oxide thin-film electrodes for electrochemical capacitor applications. J Electrochem Soc 153:A383. doi:10.1149/1.2147406

Kim Y-T, Tadai K, Mitani T (2005) Highly dispersed ruthenium oxide nanoparticles on carboxylated carbon nanotubes for supercapacitor electrode materials. J Mater Chem 15:4914–4921. doi:10.1039/B511869G

Kim DJ, Kim JK, Lee JH et al (2016) Scalable and bendable organized mesoporous TiN films templated by using a dual-functional amphiphilic graft copolymer for solid supercapacitors. J Mater Chem A 4:12497–12503. doi:10.1039/C6TA03475F

Kong L, Zhang C, Zhang S et al (2014) High-power and high-energy asymmetric supercapacitors based on Li+-intercalation into a T-Nb_2O_5/graphene pseudocapacitive electrode. J Mater Chem A 2:17962–17970. doi:10.1039/C4TA03604B

Kong L, Zhang C, Wang J et al (2015) Free-standing T-Nb_2O_5/graphene composite papers with ultrahigh gravimetric/volumetric capacitance for Li-Ion intercalation pseudocapacitor. ACS Nano 9:11200–11208. doi:10.1021/acsnano.5b04737

Laforgue A, Simon P, Sarrazin C, Fauvarque J-F (1999) Polythiophene-based supercapacitors. J Power Sour 80:142–148. doi:10.1016/S0378-7753(98)00258-4

Laforgue A, Simon P, Fauvarque JF et al (2003) Activated carbon/conducting polymer hybrid supercapacitors. J Electrochem Soc 150:A645–A651. doi:10.1149/1.1566411

Lee HY, Goodenough JB (1999) Supercapacitor behavior with KCl Electrolyte. J Solid State Chem 144:220–223. doi:10.1006/jssc.1998.8128

Lee J-K, Pathan HM, Jung K-D, Joo O-S (2006a) Electrochemical capacitance of nanocomposite films formed by loading carbon nanotubes with ruthenium oxide. J Power Sour 159:1527–1531. doi:10.1016/j.jpowsour.2005.11.063

Lee J, Kim J, Hyeon T (2006b) Recent progress in the synthesis of porous carbon materials. Adv Mater 18:2073–2094. doi:10.1002/adma.200501576

Lee H, Cho MS, Kim IH et al (2010a) RuOx/polypyrrole nanocomposite electrode for electrochemical capacitors. Synth Met 160:1055–1059. doi:10.1016/j.synthmet.2010.02.026

Lee K-T, Tsai C-B, Ho W-H, Wu N-L (2010b) Superabsorbent polymer binder for achieving MnO_2 supercapacitors of greatly enhanced capacitance density. Electrochem Commun 12:886–889. doi:10.1016/j.elecom.2010.04.012

Lee H, Kim H, Cho MS et al (2011) Fabrication of polypyrrole (PPy)/carbon nanotube (CNT) composite electrode on ceramic fabric for supercapacitor applications. Electrochim Acta 56:7460–7466. doi:10.1016/j.electacta.2011.06.113

Lehtimäki S, Suominen M, Damlin P et al (2015) Preparation of supercapacitors on flexible substrates with electrodeposited PEDOT/graphene composites. ACS Appl Mater Interfaces 7:22137–22147. doi:10.1021/acsami.5b05937

Leitner K, Lerf A, Winter M et al (2006) Nomex-derived activated carbon fibers as electrode materials in carbon based supercapacitors. J Power Sour 153:419–423. doi:10.1016/j.jpowsour.2005.05.078

Li Y, Zhao Y, Cheng H et al (2012) Nitrogen-doped graphene quantum dots with oxygen-rich functional groups. J Am Chem Soc 134:15–18. doi:10.1021/ja206030c

Liao G, Geier S, Mahrholz T et al (2015) $Li_{1.4}Al_{0.4}Ti_{1.6}(PO_4)_3$ used as solid electrolyte for structural supercapacitors. V001T01A006

Lin C, Ritter JA, Popov BN (1999) Development of carbon-metal oxide supercapacitors from sol-gel derived carbon-ruthenium xerogels. J Electrochem Soc 146:3155–3160. doi:10.1149/1.1392448

Liu X, Pickup PG (2008) Performance and low temperature behaviour of hydrous ruthenium oxide supercapacitors with improved power densities. Energy Environ Sci 1:494–500. doi:10.1039/B809939A

Liu T, Pell WG, Conway BE (1997) Self-discharge and potential recovery phenomena at thermally and electrochemically prepared RuO_2 supercapacitor electrodes. Electrochim Acta 42:3541–3552. doi:10.1016/S0013-4686(97)81190-5

Liu A, Li C, Bai H, Shi G (2010) Electrochemical deposition of polypyrrole/sulfonated graphene composite films. J Phys Chem C 114:22783–22789. doi:10.1021/jp108826e

Long JW, Bélanger D, Brousse T et al (2011) Asymmetric electrochemical capacitors—stretching the limits of aqueous electrolytes. MRS Bull 36:513–522. doi:10.1557/mrs.2011.137

Lota K, Khomenko V, Frackowiak E (2004) Capacitance properties of poly(3, 4-ethylenedioxythiophene)/carbon nanotubes composites. J Phys Chem Solids 65:295–301. doi:10.1016/j.jpcs.2003.10.051

Luo PG, Sahu S, Yang S-T et al (2013) Carbon "quantum" dots for optical bioimaging. J Mater Chem B 1:2116–2127. doi:10.1039/C3TB00018D

Ma R, Bando Y, Zhang L, Sasaki T (2004) Layered MnO_2 nanobelts: hydrothermal synthesis and electrochemical measurements. Adv Mater 16:918–922. doi:10.1002/adma.200306592

Ma G, Li J, Sun K et al (2014) High performance solid-state supercapacitor with PVA–KOH–K3 [Fe(CN)6] gel polymer as electrolyte and separator. J Power Sour 256:281–287. doi:10.1016/j.jpowsour.2014.01.062

Mastragostino M, Arbizzani C, Meneghello L, Paraventi R (1996) Electronically conducting polymers and activated carbon: electrode materials in supercapacitor technology. Adv Mater 8:331–334. doi:10.1002/adma.19960080409

Mastragostino M, Paraventi R, Zanelli A (2000) Supercapacitors based on composite polymer electrodes. J Electrochem Soc 147:3167–3170. doi:10.1149/1.1393878

Misnon II, Aziz RA, Zain NKM et al (2014) High performance MnO_2 nanoflower electrode and the relationship between solvated ion size and specific capacitance in highly conductive electrolytes. Mater Res Bull 57:221–230. doi:10.1016/j.materresbull.2014.05.044

Momma T, Liu X, Osaka T et al (1996) Electrochemical modification of active carbon fiber electrode and its application to double-layer capacitor. J Power Sour 60:249–253. doi:10.1016/S0378-7753(96)80018-8

Morishita T, Soneda Y, Tsumura T, Inagaki M (2006) Preparation of porous carbons from thermoplastic precursors and their performance for electric double layer capacitors. Carbon N Y 44:2360–2367. doi:10.1016/j.carbon.2006.04.030

Mun Y, Jo C, Hyeon T et al (2013) Simple synthesis of hierarchically structured partially graphitized carbon by emulsion/block-copolymer co-template method for high power supercapacitors. Carbon N Y 64:391–402. doi:10.1016/j.carbon.2013.07.092

Nam K-W, Kim K-H, Lee E-S et al (2008) Pseudocapacitive properties of electrochemically prepared nickel oxides on 3-dimensional carbon nanotube film substrates. J Power Sour 182:642–652. doi:10.1016/j.jpowsour.2008.03.090

Naoi K, Suematsu S, Manago A (2000) Electrochemistry of poly(1, 5-diaminoanthraquinone) and its application in electrochemical capacitor materials. J Electrochem Soc 147:420–426. doi:10.1149/1.1393212

Naudin É, El Mehdi N, Soucy C et al (2001) Poly(3-arylthiophenes): syntheses of monomers and spectroscopic and electrochemical characterization of the corresponding polymers. Chem Mater 13:634–642. doi:10.1021/cm0007656

Pan D, Zhang J, Li Z, Wu M (2010) Hydrothermal route for cutting graphene sheets into blue-luminescent graphene quantum dots. Adv Mater 22:734–738. doi:10.1002/adma.200902825

Panić V, Vidaković T, Gojković S et al (2003) The properties of carbon-supported hydrous ruthenium oxide obtained from RuOxHy sol. Electrochim Acta 48:3805–3813. doi:10.1016/S0013-4686(03)00514-0

Peng J, Gao W, Gupta BK et al (2012) Graphene quantum dots derived from carbon fibers. Nano Lett 12:844–849. doi:10.1021/nl2038979

Perricone E, Chamas M, Cointeaux L et al (2013) Investigation of methoxypropionitrile as co-solvent for ethylene carbonate based electrolyte in supercapacitors. A safe and wide temperature range electrolyte. Electrochim Acta 93:1–7. doi:10.1016/j.electacta.2013.01.084

Pico F, Ibañez J, Lillo-Rodenas MA et al (2008) Understanding $RuO_2 \cdot xH_2O$/carbon nanofibre composites as supercapacitor electrodes. J Power Sour 176:417–425. doi:10.1016/j.jpowsour.2007.11.001

Pico F, Morales E, Fernandez JA et al (2009) Ruthenium oxide/carbon composites with microporous or mesoporous carbon as support and prepared by two procedures. A comparative study as supercapacitor electrodes. Electrochim Acta 54:2239–2245. doi:10.1016/j.electacta.2008.10.028

Prasad KR, Koga K, Miura N (2004) Electrochemical deposition of nanostructured indium oxide: high-performance electrode material for redox supercapacitors. Chem Mater 16:1845–1847. doi:10.1021/cm0497576

Prasad KS, Pallela R, Kim DM, Shim YB (2013) Microwave-assisted one-pot synthesis of metal-free nitrogen and phosphorus dual-doped nanocarbon for electrocatalysis and cell imaging. Part Part Syst Charact 30:557–564. doi:10.1002/ppsc.201300020

Qu D, Shi H (1998) Studies of activated carbons used in double-layer capacitors. J Power Sour 74:99–107. doi:10.1016/S0378-7753(98)00038-X

Raccichini R, Varzi A, Passerini S, Scrosati B (2015) The role of graphene for electrochemical energy storage. Nat Mater 14:271–279

Rajendra Prasad K, Miura N (2004) Electrochemically synthesized MnO_2-based mixed oxides for high performance redox supercapacitors. Electrochem Commun 6:1004–1008. doi:10.1016/j.elecom.2004.07.017

Ramani M, Haran BS, White RE et al (2001) Studies on activated carbon capacitor materials loaded with different amounts of ruthenium oxide. J Power Sources 93:209–214. doi:10.1016/S0378-7753(00)00575-9

Ramesh TN, Kamath PV, Shivakumara C (2005) Correlation of structural disorder with the reversible discharge capacity of nickel hydroxide electrode. J Electrochem Soc 152:A806–A810. doi:10.1149/1.1865852

Ratajczak P, Jurewicz K, Skowron P et al (2014) Effect of accelerated ageing on the performance of high voltage carbon/carbon electrochemical capacitors in salt aqueous electrolyte. Electrochim Acta 130:344–350. doi:10.1016/j.electacta.2014.02.140

Ratha S, Samantara AK, Singha KK et al (2017) Urea-assisted room temperature stabilized metastable β-NiMoO$_4$: experimental and theoretical insights into its unique bifunctional activity toward oxygen evolution and supercapacitor. ACS Appl Mater Interfaces 9:9640–9653. doi:10.1021/acsami.6b16250

Raymundo-Piñero E, Cazorla-Amorós D, Linares-Solano A et al (2002) High surface area carbon nanotubes prepared by chemical activation. Carbon N Y 40:1614–1617. doi:10.1016/S0008-6223(02)00134-3

Raymundo-Piñero E, Kierzek K, Machnikowski J, Béguin F (2006) Relationship between the nanoporous texture of activated carbons and their capacitance properties in different electrolytes. Carbon N Y 44:2498–2507. doi:10.1016/j.carbon.2006.05.022

Regisser F, Lavoie M-A, Champagne GY, Bélanger D (1996) Randomly oriented graphite electrode. Part 1. Effect of electrochemical pretreatment on the electrochemical behavior and chemical composition of the electrode. J Electroanal Chem 415:47–54. doi:10.1016/S0022-0728(96)04636-0

Rudge A, Davey J, Raistrick I et al (1994) Conducting polymers as active materials in electrochemical capacitors. J Power Sour 47:89–107. doi:10.1016/0378-7753(94)80053-7

Ryu KS, Kim KM, Park N-G et al (2002) Symmetric redox supercapacitor with conducting polyaniline electrodes. J Power Sources 103:305–309. doi:10.1016/S0378-7753(01)00862-X

Ryu I, Yang M, Kwon H et al (2014) Coaxial RuO$_2$–ITO nanopillars for transparent supercapacitor application. Langmuir 30:1704–1709. doi:10.1021/la4044599

Sakiyama K, Onishi S, Ishihara K et al (1993) Deposition and properties of reactively sputtered ruthenium dioxide films. J Electrochem Soc 140:834–839. doi:10.1149/1.2056168

Salitra G, Soffer A, Eliad L et al (2000) Carbon electrodes for double-layer capacitors I. Relations between ion and pore dimensions. J Electrochem Soc 147:2486–2493. doi:10.1149/1.1393557

Samantara AK, Chandra Sahu S, Ghosh A, Jena BK (2015) Sandwiched graphene with nitrogen, sulphur co-doped CQDs: an efficient metal-free material for energy storage and conversion applications. J Mater Chem A 3:16961–16970. doi:10.1039/C5TA03376D

Samantara AK, Maji S, Ghosh A et al (2016) Good's buffer derived highly emissive carbon quantum dots: excellent biocompatible anticancer drug carrier. J Mater Chem B 4:2412–2420. doi:10.1039/C6TB00081A

Sarangapani S, Tilak BV, Chen C-P (1996) Materials for electrochemical capacitors: theoretical and experimental constraints. J Electrochem Soc 143:3791–3799. doi:10.1149/1.1837291

Seredych M, Hulicova-Jurcakova D, Lu GQ, Bandosz TJ (2008) Surface functional groups of carbons and the effects of their chemical character, density and accessibility to ions on electrochemical performance. Carbon N Y 46:1475–1488. doi:10.1016/j.carbon.2008.06.027

Sevilla M, Fuertes AB (2014) Direct synthesis of highly porous interconnected carbon nanosheets and their application as high-performance supercapacitors. ACS Nano 8:5069–5078. doi:10.1021/nn501124h

Sharma P, Bhatti TS (2010) A review on electrochemical double-layer capacitors. Energy Convers Manag 51:2901–2912. doi:10.1016/j.enconman.2010.06.031

Sharma RK, Rastogi AC, Desu SB (2008) Manganese oxide embedded polypyrrole nanocomposites for electrochemical supercapacitor. Electrochim Acta 53:7690–7695. doi:10.1016/j.electacta.2008.04.028

Shen L, Zhang L, Chen M et al (2013) The production of pH-sensitive photoluminescent carbon nanoparticles by the carbonization of polyethylenimine and their use for bioimaging. Carbon N Y 55:343–349. doi:10.1016/j.carbon.2012.12.074

Shi M, Kou S, Yan X (2014a) Engineering the electrochemical capacitive properties of graphene sheets in ionic-liquid electrolytes by correct selection of anions. Chemsuschem 7:3053–3062. doi:10.1002/cssc.201402275

Shi Y, Pan L, Liu B et al (2014b) Nanostructured conductive polypyrrole hydrogels as high-performance, flexible supercapacitor electrodes. J Mater Chem A 2:6086–6091. doi:10. 1039/C4TA00484A

Shimizu W, Makino S, Takahashi K et al (2013) Development of a 4.2 V aqueous hybrid electrochemical capacitor based on MnO_2 positive and protected Li negative electrodes. J Power Sour 241:572–577. doi:10.1016/j.jpowsour.2013.05.003

Shinde DB, Pillai VK (2012) Electrochemical preparation of luminescent graphene quantum dots from multiwalled carbon nanotubes. Chem A Eur J 18:12522–12528. doi:10.1002/chem. 201201043

Shiraishi S, Kurihara H, Okabe K et al (2002) Electric double layer capacitance of highly pure single-walled carbon nanotubes (HiPco^{TM} Buckytubes^{TM}) in propylene carbonate electrolytes. Electrochem Commun 4:593–598. doi:10.1016/S1388-2481(02)00382-X

Shulga YM, Baskakov SA, Smirnov VA et al (2014) Graphene oxide films as separators of polyaniline-based supercapacitors. J Power Sour 245:33–36. doi:10.1016/j.jpowsour.2013.06. 094

Simon P, Gogotsi Y (2008) Materials for electrochemical capacitors. Nat Mater 7:845–854. doi:10. 1038/nmat2297

Sivaraman P, Thakur A, Kushwaha RK et al (2006) Poly(3-methyl thiophene)-activated carbon hybrid supercapacitor based on gel polymer electrolyte. Electrochem Solid-State Lett 9:A435–A438. doi:10.1149/1.2213357

Snook GA, Kao P, Best AS (2011) Conducting-polymer-based supercapacitor devices and electrodes. J Power Sour 196:1–12. doi:10.1016/j.jpowsour.2010.06.084

Stankovich S, Dikin DA, Dommett GHB et al (2006) Graphene-based composite materials. Nature 442:282–286

Sugimoto W, Iwata H, Murakami Y, Takasu Y (2004) Electrochemical capacitor behavior of layered ruthenic acid hydrate. J Electrochem Soc 151:A1181–A1187. doi:10.1149/1.1765681

Sugimoto W, Iwata H, Yasunaga Y et al (2003) Preparation of ruthenic acid nanosheets and utilization of its interlayer surface for electrochemical energy storage. Angew Chemie Int Ed 42:4092–4096. doi:10.1002/anie.200351691

Sugimoto W, Iwata H, Yokoshima K et al (2005) Proton and electron conductivity in hydrous ruthenium oxides evaluated by electrochemical impedance spectroscopy: the origin of large capacitance. J Phys Chem B 109:7330–7338. doi:10.1021/jp044252o

Sugimoto W, Yokoshima K, Murakami Y, Takasu Y (2006) Charge storage mechanism of nanostructured anhydrous and hydrous ruthenium-based oxides. Electrochim Acta 52.1742–1748. doi:10.1016/j.electacta.2006.02.054

Sun D, Ban R, Zhang P-H et al (2013) Hair fiber as a precursor for synthesizing of sulfur- and nitrogen-co-doped carbon dots with tunable luminescence properties. Carbon N Y 64:424–434. doi:10.1016/j.carbon.2013.07.095

Tang L, Ji R, Cao X et al (2012) Deep ultraviolet photoluminescence of water-soluble self-passivated graphene quantum dots. ACS Nano 6:5102–5110. doi:10.1021/nn300760g

Tõnurist K, Thomberg T, Jänes A et al (2012) Specific performance of electrical double layer capacitors based on different separator materials in room temperature ionic liquid. Electrochem Commun 22:77–80. doi:10.1016/j.elecom.2012.05.029

Tsay K-C, Zhang L, Zhang J (2012) Effects of electrode layer composition/thickness and electrolyte concentration on both specific capacitance and energy density of supercapacitor. Electrochim Acta 60:428–436. doi:10.1016/j.electacta.2011.11.087

Varzi A, Balducci A, Passerini S (2014) Natural cellulose: a green alternative binder for high voltage electrochemical double layer capacitors containing ionic liquid-based electrolytes. J Electrochem Soc 161:A368–A375

Villers D, Jobin D, Soucy C et al (2003) The influence of the range of electroactivity and capacitance of conducting polymers on the performance of carbon conducting polymer hybrid supercapacitor. J Electrochem Soc 150:A747–A752. doi:10.1149/1.1571530

Vol'fkovich YM, Serdyuk TM (2002) Electrochemical capacitors. Russ J Electrochem 38:935–959. doi:10.1023/A:1020220425954

Wang Y, Hu A (2014) Carbon quantum dots: synthesis, properties and applications. J Mater Chem C 2:6921–6939. doi:10.1039/C4TC00988F

Wang Y-G, Li H-Q, Xia Y-Y (2006) Ordered whiskerlike polyaniline grown on the surface of mesoporous carbon and its electrochemical capacitance performance. Adv Mater 18:2619–2623. doi:10.1002/adma.200600445

Wang G, Shen X, Horvat J et al (2009a) Hydrothermal synthesis and optical, magnetic, and supercapacitance properties of nanoporous cobalt oxide nanorods. J Phys Chem C 113:4357–4361. doi:10.1021/jp8106149

Wang Y, Shi Z, Huang Y et al (2009b) Supercapacitor devices based on graphene materials. J Phys Chem C 113:13103–13107. doi:10.1021/jp902214f

Wang X, Cao L, Yang S-T et al (2010) Bandgap-like strong fluorescence in functionalized carbon nanoparticles. Angew Chemie Int Ed 49:5310–5314. doi:10.1002/anie.201000982

Wang G, Zhang L, Zhang J (2012a) A review of electrode materials for electrochemical supercapacitors. Chem Soc Rev 41:797–828. doi:10.1039/C1CS15060J

Wang J, Wang C-F, Chen S (2012b) Amphiphilic egg-derived carbon dots: rapid plasma fabrication, pyrolysis process, and multicolor printing patterns. Angew Chemie Int Ed 51:9297–9301. doi:10.1002/anie.201204381

Wang Q, Yan J, Wang Y et al (2014) Three-dimensional flower-like and hierarchical porous carbon materials as high-rate performance electrodes for supercapacitors. Carbon N Y 67:119–127. doi:10.1016/j.carbon.2013.09.070

Wee G, Soh HZ, Cheah YL et al (2010) Synthesis and electrochemical properties of electrospun V_2O_5 nanofibers as supercapacitor electrodes. J Mater Chem 20:6720–6725. doi:10.1039/C0JM00059K

Wen J, Ruan X, Zhou Z (2009) Preparation and electrochemical performance of novel ruthenium–manganese oxide electrode materials for electrochemical capacitors. J Phys Chem Solids 70:816–820. doi:10.1016/j.jpcs.2009.03.015

Wu Q, Xu Y, Yao Z et al (2010a) Supercapacitors based on flexible graphene/polyaniline nanofiber composite films. ACS Nano 4:1963–1970. doi:10.1021/nn1000035

Wu Z-S, Wang D-W, Ren W et al (2010b) Anchoring hydrous RuO_2 on graphene sheets for high-performance electrochemical capacitors. Adv Funct Mater 20:3595–3602. doi:10.1002/adfm.201001054

Xia H, Feng J, Wang H et al (2010) MnO_2 nanotube and nanowire arrays by electrochemical deposition for supercapacitors. J Power Sour 195:4410–4413. doi:10.1016/j.jpowsour.2010.01.075

Xia X, Tu J, Mai Y et al (2011) Graphene sheet/porous nio hybrid film for supercapacitor applications. Chem—A Eur J 17:10898–10905 doi:10.1002/chem.201100727

Xiang F, Zhong J, Gu N et al (2014) Far-infrared reduced graphene oxide as high performance electrodes for supercapacitors. Carbon N Y 75:201–208. doi:10.1016/j.carbon.2014.03.053

Xiao J, Yang S, Wan L et al (2014) Electrodeposition of manganese oxide nanosheets on a continuous three-dimensional nickel porous scaffold for high performance electrochemical capacitors. J Power Sour 245:1027–1034. doi:10.1016/j.jpowsour.2013.07.024

Xing W, Qiao S, Wu X et al (2011) Exaggerated capacitance using electrochemically active nickel foam as current collector in electrochemical measurement. J Power Sour 196:4123–4127. doi:10.1016/j.jpowsour.2010.12.003

Xu X, Ray R, Gu Y et al (2004) Electrophoretic analysis and purification of fluorescent single-walled carbon nanotube fragments. J Am Chem Soc 126:12736–12737. doi:10.1021/ja040082h

Xu B, Wu F, Chen R et al (2008) Highly mesoporous and high surface area carbon: a high capacitance electrode material for EDLCs with various electrolytes. Electrochem Commun 10:795–797. doi:10.1016/j.elecom.2008.02.033

Xu J, Wang Q, Wang X et al (2013) Flexible asymmetric supercapacitors based upon Co9S8 Nanorod//Co_3O_4@RuO_2 nanosheet arrays on carbon cloth. ACS Nano 7:5453–5462. doi:10.1021/nn401450s

Yan J, Wei T, Shao B et al (2010) Electrochemical properties of graphene nanosheet/carbon black composites as electrodes for supercapacitors. Carbon N Y 48:1731–1737. doi:10.1016/j. carbon.2010.01.014

Yan J, Wang Q, Wei T, Fan Z (2014) Recent advances in design and fabrication of electrochemical supercapacitors with high energy densities. Adv Energy Mater 4:1300816–n/a. doi:10.1002/aenm.201300816

Yang Y, Cui J, Zheng M et al (2012) One-step synthesis of amino-functionalized fluorescent carbon nanoparticles by hydrothermal carbonization of chitosan. Chem Commun 48:380–382. doi:10.1039/C1CC15678K

Yokoshima K, Shibutani T, Hirota M et al (2006) Electrochemical supercapacitor behavior of nanoparticulate rutile-type $Ru_{1-x}V_xO_2$. J Power Sour 160:1480–1486. doi:10.1016/j.jpowsour.2006.02.053

Yu G-Y, Chen W-X, Zheng Y-F et al (2006) Synthesis of Ru/carbon nanocomposites by polyol process for electrochemical supercapacitor electrodes. Mater Lett 60:2453–2456. doi:10.1016/j.matlet.2006.01.015

Yu G, Hu L, Vosgueritchian M et al (2011) Solution-processed graphene/MnO_2 nanostructured textiles for high-performance electrochemical capacitors. Nano Lett 11:2905–2911. doi:10.1021/nl2013828

Yu H, Tang Q, Wu J et al (2012) Using eggshell membrane as a separator in supercapacitor. J Power Sour 206:463–468. doi:10.1016/j.jpowsour.2012.01.116

Yuan J, Liu Z-H, Qiao S et al (2009) Fabrication of MnO_2-pillared layered manganese oxide through an exfoliation/reassembling and oxidation process. J Power Sour 189:1278–1283. doi:10.1016/j.jpowsour.2008.12.148

Zhai X, Zhang P, Liu C et al (2012) Highly luminescent carbon nanodots by microwave-assisted pyrolysis. Chem Commun 48:7955–7957. doi:10.1039/C2CC33869F

Zhang SW, Chen GZ (2008) Manganese oxide based materials for supercapacitors. Energy Mater 3:186–200. doi:10.1179/174892409X427940

Zhang H, Cao G, Wang Z et al (2008) Tube-covering-tube nanostructured polyaniline/carbon nanotube array composite electrode with high capacitance and superior rate performance as well as good cycling stability. Electrochem Commun 10:1056–1059. doi:10.1016/j.elecom.2008.05.007

Zhang Y, Feng H, Wu X et al (2009) Progress of electrochemical capacitor electrode materials: a review. Int J Hydrogen Energy 34:4889–4899. doi:10.1016/j.ijhydene.2009.04.005

Zhang K, Ang BT, Zhang LL et al (2011) Pyrolyzed graphene oxide/resorcinol-formaldehyde resin composites as high-performance supercapacitor electrodes. J Mater Chem 21:2663–2670. doi:10.1039/C0JM02850A

Zhao D-D, Bao S-J, Zhou W-J, Li H-L (2007) Preparation of hexagonal nanoporous nickel hydroxide film and its application for electrochemical capacitor. Electrochem Commun 9:869–874. doi:10.1016/j.elecom.2006.11.030

Zheng JP, Cygan PJ, Jow TR (1995) Hydrous ruthenium oxide as an electrode material for electrochemical capacitors. J Electrochem Soc 142:2699–2703. doi:10.1149/1.2050077

Zheng C, Yoshio M, Qi L, Wang H (2014) A 4 V-electrochemical capacitor using electrode and electrolyte materials free of metals. J Power Sour 260:19–26. doi:10.1016/j.jpowsour.2014.02.098

Zhong C, Deng Y, Hu W et al (2015a) A review of electrolyte materials and compositions for electrochemical supercapacitors. Chem Soc Rev 44:7484–7539. doi:10.1039/C5CS00303B

Zhong J, Fan L-Q, Wu X et al (2015b) Improved energy density of quasi-solid-state supercapacitors using sandwich-type redox-active gel polymer electrolytes. Electrochim Acta 166:150–156. doi:10.1016/j.electacta.2015.03.114

Zhu S, Zhang J, Qiao C et al (2011) Strongly green-photoluminescent graphene quantum dots for bioimaging applications. Chem Commun 47:6858–6860. doi:10.1039/C1CC11122A

Chapter 4
Asymmetric and Hybrid Supercapacitor

Both EDLCs and pseudocapacitor produce enough power density easily surpassing that of a battery, but their energy density is too low which is why they are currently employed only as backup devices for the purpose of rapid storage and discharge process. There are attempts made to combine electrodes of contrasting properties in order to achieve higher operating potential and to prevent electrolytic decomposition. For example, a strategic combination of carbonaceous material based electrode (EDLC electrode) and another electrode made out of pseudocapacitive material could exploit the altogether different potential windows so that the total working potential is the resultant of their individual working potential window optimized via suitable electrochemical method. Such tactical manipulation in selecting two asymmetric electrodes is highly essential. The primary goal is to devise a supercapacitor that would possess desired energy density (due to the presence of pseudocapacitive material) without sacrificing its characteristic power density (maintained by the EDLC type electrode).

Furthermore, approach has been made to combine both the basic principles of battery and supercapacitor to fabricate a device that could emulate the exceptional energy density of a battery and unmatched power density of a supercapacitor. Such devices are called as hybrid supercapacitors or more appropriately battery-supercapacitor hybrid devices (BSH) (Li et al. 2015a, b). Unlike electric double layer supercapacitor (EDLS) and pseudocapacitors, a BSH makes use of an EDLC material (activated carbon/grapheme/CNTs etc.) as the cathode and a suitable battery electrode as the anode. While the battery electrode (anode) acts as the energy source, the cathode (capacitor material) operates as the power source so as to replace the battery technology. Such an assembly of two dissimilar electrodes is meant to bring a balance between both power and energy density. Similar to asymmetric supercapacitor (ASC) devices, electrodes in the hybrid supercapacitors operate at different potential windows. Taking advantage of this unique property, the overall potential window can be escalated to attain higher values which in turn will enhance the energy density without hindering the power density. Therefore, to

© The Author(s) 2018   41
A. K. Samantara and S. Ratha, *Materials Development for Active/Passive
Components of a Supercapacitor*, SpringerBriefs in Materials,
https://doi.org/10.1007/978-981-10-7263-5_4

achieve high energy and power density, the designing and development of suitable electrode materials is highly essential.

However, the selection and fabrication process is not as simple as is observed in the case of a typical EDLC and/or pseudocapacitor device. First, the optimum working potential window is achieved for both the electrodes using a standard three-electrode electrochemical cell (half-cell arrangement). From the half-cell measurements, details such as the loaded mass, capacitance value, and potential window are calculated and used to evaluate the required mass for each of the electrode material so as to maintain the mass-charge balance which is critical for the stability and efficient operation of the ASC device. Following equation is used to evaluate the mass ratio of the two electrodes (Shen et al. 2015; Wu et al. 2017):

$$\frac{m_+}{m_-} = \frac{C_- \Delta E_-}{C_+ \Delta E_+} \tag{1}$$

where m, C, and ΔE are the electrode mass, supercapacitor, and the potential range in the charge/discharge process for both electrodes, respectively.

4.1 Asymmetric Supercapacitor (ASC)

ASCs consist of a cathode, fabricated using a carbonaceous material having high specific surface area (SSA) and an anode comprising high performance and cost effective pseudocapacitor materials such as metal oxides (MnO_2, Fe_3O_4, RuO_2, MoO_3, PbO_2 and Fe_2O_3 etc.), conducting polymers such as PANI, PPy etc. (Zhi et al. 2013; Grote and Lei 2014; Long et al. 2014; Zhou et al. 2014a, b; Liu et al. 2015; Wu et al. 2017). The strategically tuned potential window in such ASCs can go beyond the decomposition potential of water (in the case of aqueous electrolytes) which clearly gives significant advantage over the symmetric supercapacitor devices having lower potential window (~ 1.0 V) in the presence of aqueous electrolytes. Apart from the compounds mentioned above, metal sulfides such as NiS (Guan et al. 2017), CoS (Shi et al. 2015), $NiCO_2S_4$ etc. (Shen et al. 2015) have shown promising results as anode materials for ASC devices. Though the cycle life of an ASC device is lower than that of EDLC type supercapacitor, it effortlessly makes up with its exceptionally balanced power and energy density. Following Fig. 4.1 is a schematic of a typical asymmetric supercapacitor device.

4.2 Hybrid Supercapacitor

Considering the current development of the EES devices, Li-ion batteries and supercapacitors are the most sought after devices which boast promising potential to be implemented in wide range of practical applications. The concept of lithium

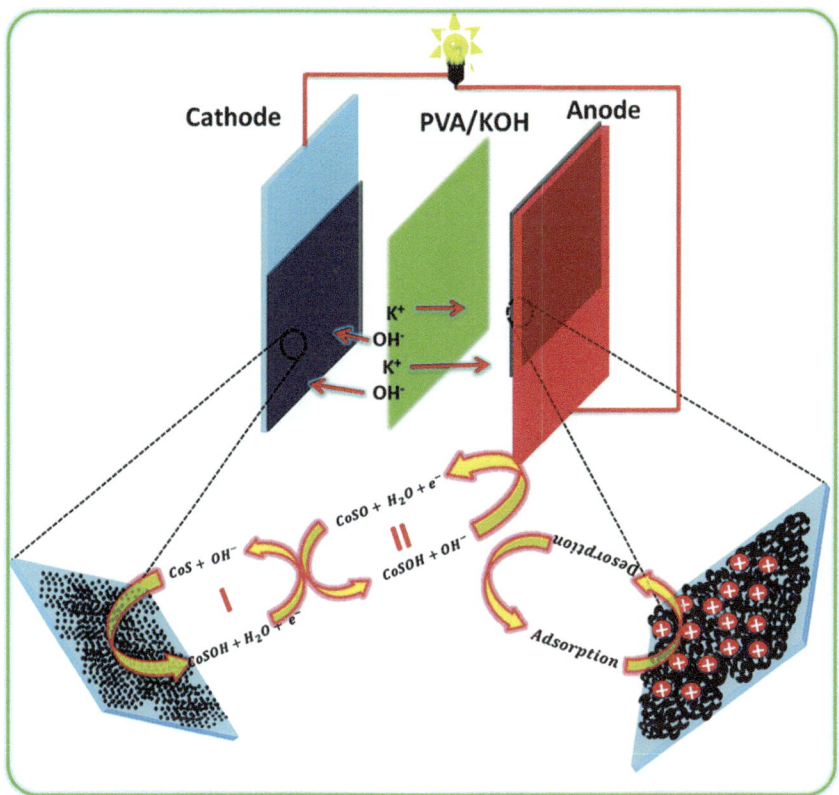

Fig. 4.1 Schematic showing an asymmetric supercapacitor device consisting of a gel polymer electrolyte (PVA/KOH). As can be seen, two different electrode materials are sandwiching the GPE in between them. Reproduced with permission from Subramani et al. (2017)

storage is promising considering the excellent electrochemical activity of Li, its abundance and it being lightweight in nature. There are numerous interesting literature reports on the high yielding lithium storage techniques which are essential in improving the LIB technology further. Efforts have been made to combine a material that has high lithium storage property with a capacitor material in order to derive high energy density like that of a Li-ion battery, still retaining the power density, typical of a supercapacitor. The Li-ion capacitor consists of a positive electrode (anode) made of capacitor material (carbonaceous) and a cathode used in typical Li-ion batteries such as $LiMn_2O_4$, $Li_4Ti_5O_{12}$, $MnFe_2O_4$, and $LiTi_2(PO_4)_3$ etc. (Aravindan et al. 2014; Yan et al. 2014). Sodium-ion hybrid capacitors have also emerged as potential energy storage devices recently (Senthilkumar et al. 2015). Figure 4.2 illustrates a basic principle behind the fabrication of a sodium-ion hybrid capacitor. These hybrid devices come with some promising specifications such as,

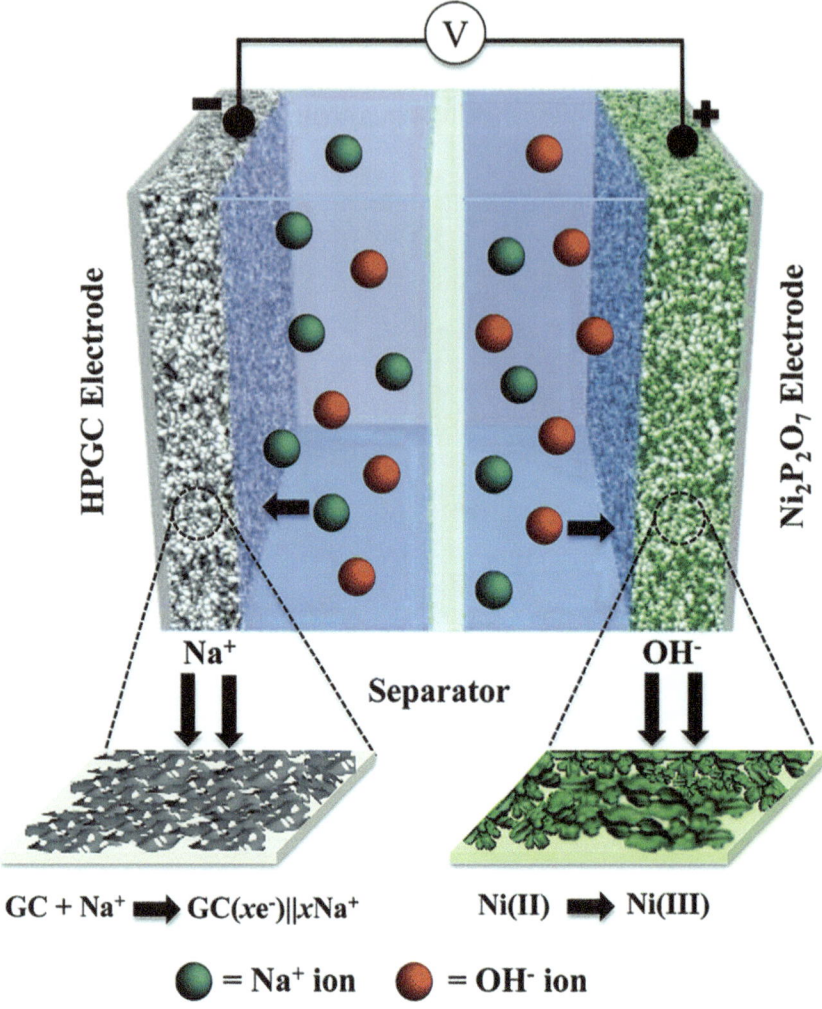

Fig. 4.2 Schematic elucidating basic mechanism behind the sodium-ion hybrid capacitor device. Reproduced with permission from Senthilkumar et al. (2015)

    (i)   Low self-discharge,
  (ii) High reliability,
 (iii) Wide operating temperature,
 (iv) Long cycle life and
  (v) High degree of safety.

These advantageous properties are gradually being exploited in various areas such as automatic guided vehicles, stacker cranes, construction machines, and instantaneous potential drop compensators.

References

Aravindan V, Gnanaraj J, Lee Y-S, Madhavi S (2014) Insertion-type electrodes for nonaqueous li-ion capacitors. Chem Rev 114:11619–11635. doi:10.1021/cr5000915

Grote F, Lei Y (2014) A complete three-dimensionally nanostructured asymmetric supercapacitor with high operating voltage window based on PPy and MnO_2. Nano Energy 10:63–70. doi:10.1016/j.nanoen.2014.08.019

Guan B, Li Y, Yin B et al (2017) Synthesis of hierarchical NiS microflowers for high performance asymmetric supercapacitor. Chem Eng J 308:1165–1173. doi:10.1016/j.cej.2016.10.016

Li Q, Chen W, Liu Z et al (2015a) Development of energy management system based on a power sharing strategy for a fuel cell-battery-supercapacitor hybrid tramway. J Power Sour 279:267–280. doi:10.1016/j.jpowsour.2014.12.042

Li R, Wang Y, Zhou C et al (2015b) Carbon-stabilized high-capacity ferroferric oxide nanorod array for flexible solid-state alkaline battery–supercapacitor hybrid device with high environmental suitability. Adv Funct Mater 25:5384–5394. doi:10.1002/adfm.201502265

Liu W, Liu N, Shi Y et al (2015) A wire-shaped flexible asymmetric supercapacitor based on carbon fiber coated with a metal oxide and a polymer. J Mater Chem A 3:13461–13467. doi:10.1039/C5TA01105A

Long C, Qi D, Wei T et al (2014) Nitrogen-doped carbon networks for high energy density supercapacitors derived from polyaniline coated bacterial cellulose. Adv Funct Mater 24:3953–3961. doi:10.1002/adfm.201304269

Senthilkumar B, Khan Z, Park S et al (2015) Highly porous graphitic carbon and $Ni_2P_2O_7$ for a high performance aqueous hybrid supercapacitor. J Mater Chem A 3:21553–21561. doi:10.1039/C5TA04737D

Shen L, Wang J, Xu G et al (2015) $NiCo_2S_4$ nanosheets grown on nitrogen-doped carbon foams as an advanced electrode for supercapacitors. Adv Energy Mater 5:n/a–n/a. doi:10.1002/aenm.201400977

Shi J, Li X, He G et al (2015) Electrodeposition of high-capacitance 3D CoS/graphene nanosheets on nickel foam for high-performance aqueous asymmetric supercapacitors. J Mater Chem A 3:20619–20626. doi:10.1039/C5TA04464B

Subramani K, Sudhan N, Divya R, Sathish M (2017) All-solid-state asymmetric supercapacitors based on cobalt hexacyanoferrate-derived CoS and activated carbon. RSC Adv 7:6648–6659. doi:10.1039/C6RA27331A

Wu L, Hao L, Pang B et al (2017) MnO_2 nanoflowers and polyaniline nanoribbons grown on hybrid graphene/Ni 3D scaffolds by in situ electrochemical techniques for high-performance asymmetric supercapacitors. J Mater Chem A 5:4629–4637. doi:10.1039/C6TA10757E

Yan J, Wang Q, Wei T, Fan Z (2014) Recent advances in design and fabrication of electrochemical supercapacitors with high energy densities. Adv Energy Mater 4:1300816–n/a. doi:10.1002/aenm.201300816

Zhi M, Xiang C, Li J et al (2013) Nanostructured carbon-metal oxide composite electrodes for supercapacitors: a review. Nanoscale 5:72–88. doi:10.1039/C2NR32040A

Zhou Y, Lachman N, Ghaffari M et al (2014a) A high performance hybrid asymmetric supercapacitor via nano-scale morphology control of graphene, conducting polymer, and carbon nanotube electrodes. J Mater Chem A 2:9964–9969. doi:10.1039/C4TA01785D

Zhou Y, Xu H, Lachman N et al (2014b) Advanced asymmetric supercapacitor based on conducting polymer and aligned carbon nanotubes with controlled nanomorphology. Nano Energy 9:176–185. doi:10.1016/j.nanoen.2014.07.007

Chapter 5
Trend and Scope Beyond Traditional Supercapacitors

Primitive type supercapacitors that were made of pure carbon or carbon derived electrode materials have gone through rapid evolution since few decades. Most of these traditional supercapacitor devices are of symmetric nature, i.e. both electrodes are fabricated from the same material which reduces any chances of enhancing the working potential window. The first stage evolution took place with the concept of taking two different materials as electrodes to form an asymmetric supercapacitor device which not only will provide the option for wider potential window but also suppress decomposition of electrolytic content (specifically for aqueous based electrolytes). Thus asymmetric supercapacitors are promising storage options and should be put to further research in order to promote its commercial viability. Extensive research should also be done on electrolytic materials as they play a key role in the performance of a supercapacitor device.

Apart from the above asymmetric supercapacitor prototype, further work need to be carried out to try and test the feasibility of various other cost-effective materials as supercapacitor electrodes. As discussed in the previous sections, there are challenges while trying to obtain a system of two electrodes using a suitable electrolytic material. There are other aspects too such as the post-fabrication performance of the supercapacitor device, safety, reliability and stability. Also, more attention is to be devoted towards the development and practical implementation of hybrid supercapacitors by designing a combination of electrodes and electrolyte.

To create a high performance asymmetric and/or hybrid storage device (supercapacitor), there are few key points which should be taken care of;

1. Designing a current collector which should be able to provide excellent contact formation. It should be highly resistant toward any physical and/or chemical hazards. Preferably, a highly conducting, robust, non-corrosive, and flexible material such as metal foam/foil is to be applied as the current collector.
2. Suitable electrode materials for the purpose to fabricate both cathode and anode. For the cathode, a material with high porosity, chemical stability, mechanical/

© The Author(s) 2018
A. K. Samantara and S. Ratha, *Materials Development for Active/Passive Components of a Supercapacitor*, SpringerBriefs in Materials,
https://doi.org/10.1007/978-981-10-7263-5_5

tensile strength, superior adsorption capability will be ideal in order to promote EDLC.

In contrast, for the anode, a material with high pseudocapacitive/lithium intercalation property, high electrical conductivity would suffice. For ASCs, though metal oxides provide plenty of options to try out, research should be extended toward more promising materials such as metal nitrides having better electrical conductivities which would have significant impact on the supercapacitor performances.

Similar approach toward the futuristic development of lithium storage material is the key to the designing of high power density and energy density supercapacitors which could assist the battery technology to ward off issues of sluggish charge uptake time, overcharging etc. Lithiated metal nitrides and/or lithium compounds are found to be ideal for this purpose.

Furthermore, electrode material synthesis is an important aspect which should be carried out with utmost care. Most of the materials that have been grown indirectly using substrate/template assisted growth can be successfully replaced with those grown directly using various substrates and can directly be taken as current collector/electrode assembly. That would bring much better adhesion between the both which is desired for a supercapacitor device.

3. Judicious selection of electrolyte materials (liquid, gel or solid) is also critical as it would have significant impact on the stability and ionic conductivity of the solid electrolyte interphase (SEI). The compactness, rigidity and long cycle life of a supercapacitor device will be determined by the type of electrolyte used. For supercapacitors employing liquid electrolytes, a suitable separator should be selected to facilitate ion/charge transportation and for maintaining a barrier between the two electrodes and protect the device from short circuit.

4. Though rigorous research has been done on huge number of metal oxides as cheap and high performance pseudocapacitive material, extended interest should be vested in few promising compounds such as metal nitrides, metal borides etc. which would facilitate the selection and development of efficient high performance supercapacitor electrodes.

GPSR Compliance

*The European Union's (EU) General Product Safety Regulation (GPSR)
is a set of rules that requires consumer products to be safe and our
obligations to ensure this.*

*If you have any concerns about our products, you can contact us on
ProductSafety@springernature.com*

In case Publisher is established outside the EU, the EU authorized
representative is:

Springer Nature Customer Service Center GmbH
Europaplatz 3
69115 Heidelberg, Germany

Batch number: 09473857

Printed by Printforce, the Netherlands